Applying Algebra from A to Z

by Margaret Thomas

illustrated by Tony Waters

cover by Jeff Van Kanegan

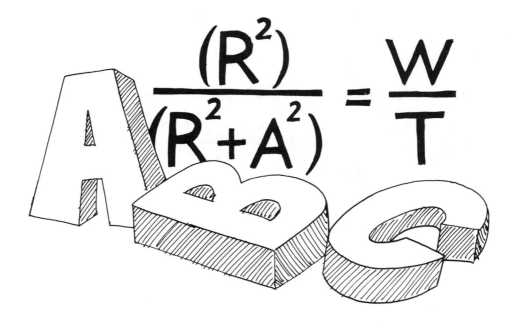

$$\frac{(R^2)}{(R^2+A^2)} = \frac{W}{T}$$

Publisher

Instructional Fair • TS Denison

Grand Rapids, Michigan 49544

ISBN: 1-56822-844-9
Applying Algebra from A to Z
Copyright © 1999 by Ideal • Instructional Fair Publishing Group
a division of Tribune Education
2400 Turner Avenue NW
Grand Rapids, MI 49544

Table of Contents

Introduction

Applying Algebra from A to Z provides the teacher with 44 math applications based on a wide range of subjects that are presented in alphabetical order. Applications are interdisciplinary in nature and draw from the fields of science, history, music, and the arts. Topics range from atoms to zero temperature, from baseball to heart disease, from calories to wave motion, and from magic to violins. The activities require a variety of math concepts including number theory, collecting data, evaluating expressions, plotting points, solving equations, ratio, proportion, and percent. Several topics include Fascinating Facts to help motivate the students.

In addition to improving algebra and calculator skills, better communication skills are also enhanced. The information presented may be used to stimulate entire class discussions, or the activities may be completed by individuals or be assigned to partners or cooperative learning groups.

Each application includes a teacher information sheet and a blackline master of a student activity sheet or a small-group task card. For each application, the teacher information sheet lists the math topics involved, includes a description of the activity, and presents background information for the teacher's use in class discussion. Some of the teacher information sheets include math shortcuts and fascinating facts. The student activity sheet includes basic information on the topic plus associated math problems. Some of the activity sheets are self-checking with correct answers resulting in the spelling of a name or phrase. The student activity sheets can be used to make copies for students to use individually or can be used as a transparency for lively class discussions. The task cards are intended for small group or partner work. All task cards list needed materials, define the task, and provide the steps involved in the activity. Applications involving task cards lend themselves well to student discussion as well as consensus of opinions and results. Cards can also be used as blackline copymasters for transparencies to be used by the teacher for classwide discussion.

Throughout the book, students are exposed to a number of different mathematical concepts. The applications of those concepts are used to demonstrate the real world use of mathematics.

Atoms

Student Activity: "Amazing Atoms"—Students read information pertaining to atoms and write the numbers in scientific notation.

Background

An *atom* is the smallest particle of an element that has the same properties as the element. The name comes from the Greek *atomos* meaning "cannot be split." An atom contains subatomic particles of protons (positive charge), electrons (negative charge), and neutrons (neutral charge). The structure of an atom consists of a core called the *nucleus*. The nucleus contains the protons and any neutrons. It is surrounded by an *electron cloud* which contains the electrons. The electrons orbit the nucleus. The number of electrons equals the number of protons, so the net charge of a stable atom is zero.

Numbers used to describe extremely small or large quantities are often written in *scientific notation*. A number in scientific notation is the product of two factors: (a number greater than or equal to 1 that is less than 10) x (a power of 10).

Example: $0.0000000067 = 6.7 \times 0.000000001 = 6.7 \times 10^{-9}$

$93,000,000 = 9.3 \times 10,000,000 = 9.3 \times 10^{7}$

Shortcut:

To convert a standard number to scientific notation: The number of places the decimal point is moved is the exponent. The exponent is negative for small decimal numbers (movement of the decimal to the right) and positive for large numbers (movement of the decimal to the left).

Fascinating Fact

If all the empty space in your body's atoms could be removed, you would be no larger than a grain of sand.

Amazing Atoms

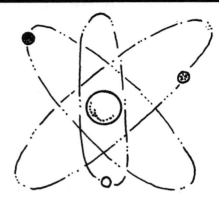

An *atom* is the smallest particle of an element that has the same properties as the element. Atoms contain protons (positive), electrons (negative), and neutrons (neutral). Numbers used to describe extremely small or large quantities are often written in *scientific notation*. A number in scientific notation is the product of two factors: (a number greater than or equal to 1 that is less than 10) x (a power of 10).

Example 1: $0.0000000067 = 6.7 \times 0.000000001 = 6.7 \times 10^{-9}$
Example 2: $93,000,000 = 9.3 \times 10,000,000 = 9.3 \times 10^{7}$

Shortcut: The number of places the decimal point is moved is the exponent. The exponent is negative for small decimal numbers (those less than 1) and positive for large numbers (those greater than 1).

Write the following numbers pertaining to atoms in scientific notation.

1. The mass of an electron is 0.00000000000000000000000091 mg. _____

2. The mass of a hydrogen atom is 0.00000000000000000000167 mg. _____

3. The diameter of a helium atom is 0.000000022 cm. _____

4. The charge of an electron is 0.000000000048 electrostatic units. _____

5. If 1,000,000 hydrogen atoms were placed side by side, it would be less than the thickness of a sheet of notebook paper. _____

6. A hydrogen atom enlarged a quadrillion times (1,000,000,000,000,000) would be a hundred miles in diameter. _____

7. If an atom were enlarged to the size of a football field, the nucleus would be about the size of a pea and would weigh 3,000,000 tons. _____

8. A standard unit of measure for elements is the *mole* (*Avogadro's number*) which equals 602,200,000,000,000,000,000,000 elements. _____

Fascinating Fact | *If all the empty space in your body's atoms could be removed, you would be no larger than a grain of sand.*

Atomic Numbers

Math Topics: Simple equations, math substitution

Student Activity: "Atomic Numbers"—Students complete a chart relating the number of protons, electrons, and neutrons in atoms and isotopes to the atomic number and atomic mass number.

Background

The *atomic number* is the number of protons in an atom. The number of protons equals the number of electrons. The net charge on a stable atom is zero. The *atomic mass number* indicates the number of protons and neutrons in the nucleus. Electrons are far less massive than protons, whereas neutrons are slightly heavier than protons. Given the atomic number and the atomic mass number of a stable atom, one can determine the number of protons (p), electrons (e), and neutrons (n).

Atomic Number = number of protons (p)
Number of protons = number of electrons (e), since the net charge is zero.
So, p = e
Atomic Mass Number = number of protons + number of neutrons (p + n)

Example:
Oxygen has an atomic number of 8 and an Atomic Mass Number of 16.
p = 8. Since p = e, e = 8.   p + n = 16, so n = 8
Oxygen has 8 protons, 8 electrons, and
8 neutrons.

All atoms of an element have the same number of protons and electrons. For example, all carbon atoms have 6 protons and 6 electrons. The number of neutrons may vary. Atoms of the same element with different numbers of neutrons are called *isotopes*. The most common isotope of carbon, Carbon 12, has 6 protons, 6 electrons, and 6 neutrons. Carbon 14 has 6 protons, 6 electrons, and 8 neutrons. The isotope Carbon 14 is used in carbon dating to determine the age of prehistoric organisms.

> **Fascinating Fact**
>
> *Laboratory-created isotopes tend to be very unstable. Copper 66 (37 neutrons) only lasts for a few minutes.*

Atomic Numbers

The *atomic number* of an element gives the number of protons in the nucleus of the atom. The number of protons equals the number of electrons. The *atomic mass number* indicates the number of protons and neutrons in the nucleus.

All atoms of an element have the same number of protons and electrons. All carbon atoms have 6 protons and 6 electrons. However, the number of neutrons may vary. Atoms of the same element with different numbers of neutrons are called *isotopes*. Carbon 12 has 6 neutrons, whereas Carbon 14 has 8 neutrons.

Given the atomic number and the atomic mass number, you can determine the number of protons (p), electrons (e), and neutrons (n) in the atoms.

atomic number = p and p = e   atomic mass = p + n

Example: Oxygen has an atomic number of 8 and an atomic mass of 16.
p = 8, so e = 8; and since p + n = 16, n = 8.

Complete the chart for the given elements and isotopes.

ELEMENT	ATOMIC NUMBER = P	ATOMIC MASS = P + N	P	E	N
1. Barium	56	137			
2. Calcium	20				20
3. Krypton				36	47
4. Mercury		200		80	
5. Silver	47				60
6. Sodium	11	23			
7. Uranium	92	238			
8. Zinc		65		30	

Baseball–
Pythagorean Theorem

Math Topics: Ratios, squares, evaluating an expression
Calculator Use: x^2 key, parentheses

Student Activity: "Pythagorean Theorem of Baseball"—Students use a formula to predict the number of wins World Series-winning teams should have had based on the runs made and runs allowed during the season.

ACTUAL NUMBER OF WINS:	PYTHAGOREAN PREDICTION:
1997 Florida Marlins—106	114
1987 Minnesota Twins—85	79
1977 New York Yankees—100	100
1967 St. Louis Cardinals—101	98
1957 Milwaukee Braves—95	94

Additional activities could include comparing the predictions to the actual results and applying the formula to data from local teams.

Background

Bill James, author of *The Baseball Abstract* defined *sabermetrics* as "the search for objective knowledge about baseball." A formula developed from data by Bill James is that the ratio of wins to losses will equal the ratio of the squares of the runs scored to the runs allowed. A related formula is the Pythagorean Theorem of Baseball:

$$(R^2)/(R^2 + A^2) = W/T \text{ or when solved for W, } (TR^2)/(R^2 + A^2) = W$$

R is the number of runs scored, A is the number of runs allowed, W is the number of wins, and T is the total number of games played.

The formula is a result of a large sample of data. It should be noted that it is a *prediction* of a team's performance.

Pythagorean Theorem of Baseball

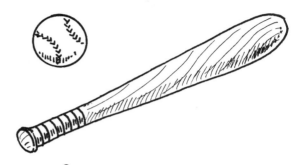

A good predictor of the number of games a team will win in a season is to consider the number of runs scored and the number of runs allowed. A formula known as the Pythagorean Theorem of Baseball models the relationship:

$$\frac{(R^2)}{(R^2 + A^2)} = \frac{W}{T} \quad \text{or} \quad \frac{(TR^2)}{(R^2 + A^2)} = W$$

R = runs scored, A = runs allowed, W = games won, T = total games played

Use the formula to predict the number of season games won by the following World Series-winning teams:

YEAR	TEAM	R	A	T	W
1997	Florida Marlins	946	749	186	1.
1987	Minnesota Twins	786	806	162	2.
1977	New York Yankees	831	651	162	3.
1967	St. Louis Cardinals	695	557	161	4.
1957	Milwaukee Braves	772	613	154	5.

Considering the data given, what is unique about the Minnesota Twins?

_____

Bouncing Balls

Math Topics: Data collecting, graphing, slope

Student Activity: "Bouncing Balls Task Card"—In small groups or partners, students drop a ball from various heights, measure the height of the bounce, and plot the data (drop height, bounce height).

Background

The height a ball bounces (b) is directly proportional to the height of the ball drop (h) so $b = kh$. The points (h, b) lie on a line. *K*, the bounce coefficient, is the slope of the line through the points (h, b).

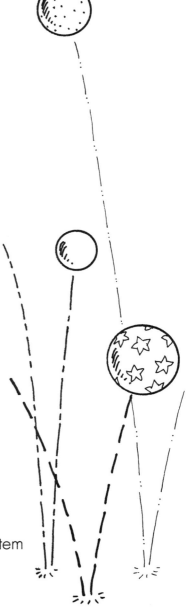

Different balls have different bounce coefficients. Vary the types of balls used by the different groups. Discuss the anticipated results. It may surprise some students to learn that no ball will bounce higher than the height of its drop. The ball is to be dropped, not thrown to the floor. Prior to being dropped, the ball has potential energy. Once the ball is released, the potential energy becomes kinetic energy. Some of the energy is transferred to the floor upon impact. The bounce height will always be less than the drop height. Therefore, $b < h$, so $k < 1$. As long as the axes have the same units, the data points (h, b) will lie on a line rising left to right at an angle less than 45° with the horizontal axis.

The activity can be used as an introduction to the concept of the slope of a line. Slope is an indication of the effect a change in the independent variable, usually x (height of the drop), has on the dependent variable, usually y (height of the drop). Slope is *the change in y divided by the change in x.*

Any direct proportion relationship can be used to demonstrate slope. Some suggested data to graph:

    (number of items, cost of items); slope equals cost of one item
    (number of feet, number of inches); slope equals 12
    (number of kg, number of lbs.); slope equals 2.2
    (number of km, number of miles); slope equals 0.6

Bouncing Balls Task Card

B

Materials:

Ball
Metric tape measure
Graph paper

 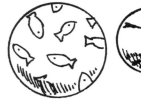

Activity:

Drop the ball from various heights (h). Record the height of the first bounce (b) for each different drop height. Measure from the bottom of the ball each time. Calculate the bounce coefficient bounce/height. Plot the points (height, bounce).

Drop Height (h):	Bounce (b):	Data Point:	Bounce Coefficient: b/h
1. 150 cm	_____	(150, _____)	_____
2. 125 cm	_____	(125, _____)	_____
3. 100 cm	_____	(100, _____)	_____
4. 75 cm	_____	(75, _____)	_____
5. 50 cm	_____	(50, _____)	_____

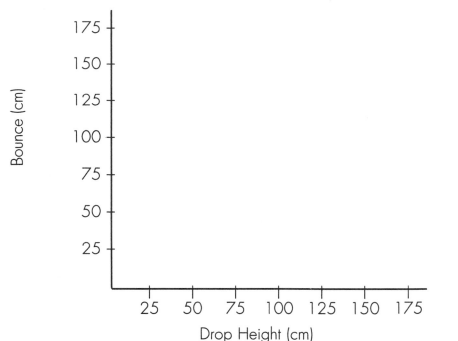

Choose two points and calculate the slope of the line:

6. Find the difference in the bounce heights. _____

7. Find the difference in the drop heights. _____

8. Divide the result from 6 by the result from 7. _____

Bowling

Student Activity: "Bowling"—Students follow a flowchart to complete bowling scorecards.

Background

A game consists of ten frames. Each frame consists of rolling one or two balls to knock down the ten pins. Scoring is cumulative. Results of each frame are shown by the following symbols and scoring follows three basic rules:

1. Strike (all pins knocked down on first ball): 10 points plus the number of pins knocked down on the next two balls rolled.

3 2. Spare (all pins knocked down on two balls): 10 points plus the number of pins knocked down on the next ball.

4 5 3. Open frame (total pins knocked down fewer than 10): add the total pins knocked down by both balls.

NOTE: If a strike or spare is made in the tenth frame, additional balls are rolled to complete the score for the frame.

Scoring flowchart:

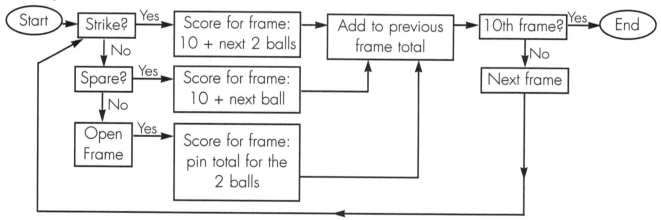

Fascinating Fact

Bowling equipment, balls and pins, were discovered in a child's grave in Egypt dating back to 5200 B.C. In the third century, Germans rolled stones at nine pins. The Dutch brought the game to America in the 1600s. After people started betting on games, Connecticut outlawed nine-pin bowling. To circumvent the law, the residents added a tenth pin. This began the ten-pin game as it is played today. The American Bowling Congress, ABC, was organized in 1895 to establish standardized rules and regulations. The Professional Bowlers Association, PBA, was established in 1958.

Bowling

A game consists of ten frames. Each frame consists of rolling one or two balls to knock down the ten pins. Scores for each frame are shown by the following symbols:

⊠   1. Strike: all pins knocked down on first ball.

3 ◩   2. Spare: all pins down on first and second balls.

4 [5]   3. Open frame: pin count for each of the two balls.

Use the following flowchart to complete the scorecards.

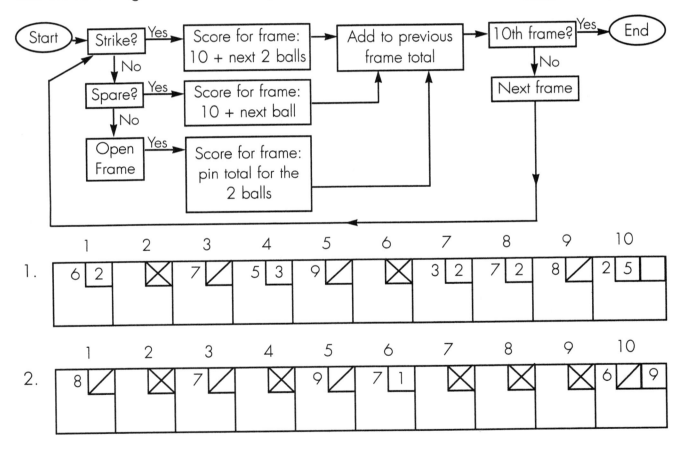

3. How can a perfect score of 300 be made? _____

Fascinating Fact	Bowling equipment, balls and pins, were discovered in a child's grave in Egypt dating back to 5200 B.C.

Calories

Math Topic: Formulas

Student Activity: "Calorie Count"—Students compute the amount of calories for food items given the grams of proteins, carbohydrates, and fat. Letters associated with the correct answers spell *Bomb Calorimeter*, which is the equipment used to measure calories.

Background

A *calorie* is a unit of heat energy. Calories describe the amount of energy foods supply. The energy is released when the digestive system breaks down the food. Three nutrients—proteins, carbohydrates, and fats—provide energy to the body. Proteins needed for amino acids that help with bone and muscle growth provide 4 calories/gram. Carbohydrates, both simple sugars and complex starches, which supply energy, provide 4 calories/gram. Fat, which helps insulate the body and protects organs, provides 9 calories/gram. Therefore, the number of calories equals $4p + 4c + 9f$, when p is the grams of proteins, c is the grams of carbohydrates, and f is the grams of fat. It is often recommended that the fat intake not exceed 30% of the daily caloric intake.

The body stores energy as fat since fat can hold more energy (1 gram = 9 calories). If more calories are taken in than are used, extra energy is stored as fat. If more calories are used than calories taken in, some stored fat is used. If the calories taken in equal the calories used, there is no change. A calorie is a calorie whether it comes from proteins, carbohydrates, or fat. Low-fat, reduced fat, and no-fat foods can result in weight gain if the number of calories consumed is greater than the calories used. Low-fat items often have increased sugar contents to maintain an acceptable taste.

Fascinating Fact

To determine the amount of calories, scientists burn food in a "Bomb Calorimeter" and measure the amount of heat produced.

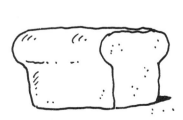

Calorie Count

Calories describe the amount of energy foods supply. Proteins (p) and carbohydrates (c) provide 4 calories/gram. Fat (f) provides 9 calories/gram.

Calories = 4p + 4c + 9f when p, c, and f are given in grams.

Calculate the calories in each serving. Place the letter of the correct answer above the problem number to discover the name of the apparatus used to measure calories.

1. Chocolate-covered peanut butter heart: protein–3 g; carbohydrate–20 g; fat–11 g

  A) 191   B) 201   C) 236

2. Hamburger sandwich: protein–13 g; carbohydrate–34 g; fat–9 g

  Q) 224   R) 269   S) 289

3. French fries: protein–6 g; carbohydrate–57 g; fat–22 g

  K) 325   L) 370   M) 450

4. Corn chips: protein–2 g; carbohydrate–15 g; fat–10 g

  R) 108   S) 138   T) 158

5. Garden cheddar tuna casserole: protein–7 g; carbohydrate–34 g; fat–4 g

  L) 200   M) 180   N) 160

6. Macaroni and cheese: protein–11 g; carbohydrate–47 g; fat–2 g

  N) 240   O) 250   P) 295

7. Oat cereal with skim milk: protein–3 g; carbohydrate–22 g; fat–2 g

  I) 118   J) 123   K) 128

8. 2% milk: protein–8 g; carbohydrate–12 g; fat–5 g

  A) 100   E) 125   I) 140

9. Crackers and cheese spread: protein–3 g; carbohydrate–9 g; fat–7 g

  B) 76   C) 111   D) 131

10. Yellow cake: protein–2 g; carbohydrate–34 g; fat–3 g

  A) 151   B) 171   C) 191

$$\overline{10}\ \overline{6}\ \overline{3}\ \overline{10}\ \ \ \ \overline{9}\ \overline{1}\ \overline{5}\ \overline{6}\ \overline{2}\ \overline{7}\ \overline{3}\ \overline{8}\ \overline{4}\ \overline{8}\ \overline{2}$$

Cricket Temperature

Student Activity: "Cricket Temperature"—Students plot data points and draw the corresponding line to answer questions concerning cricket chirps and temperature.

Background

The number of chirps per minute that some crickets make is related to the temperature. The relationship is very close to being a linear relationship.

Depending on the math background of students, an equation for the relationship can be found by using the two given points to determine the slope of the line: $m = (80 - 68)/(172 - 124)$; $m = \frac{1}{4}$. A data point and the slope can be substituted into the point-slope form of an equation to obtain the equation: $T - 80 = (\frac{1}{4})(C - 172)$ or $T = (\frac{1}{4})C + 37$, where T is the temperature in °F and C is the number of cricket chirps per minute.

Fascinating Fact

A close approximation for the Celsius temperature is $\frac{1}{8} C + 5$ where C is the number of chirps in one minute.

 Cricket Temperature

The number of chirps per minute that some crickets make is related to the temperature. The relationship is very close to being a linear relationship.

When the crickets chirped 172 times a minute, the temperature was 80°F.
Likewise, when the crickets chirped 124 times a minute, the temperature was 68°F.

Plot the information on the graph below. Draw a straight line through the two points.

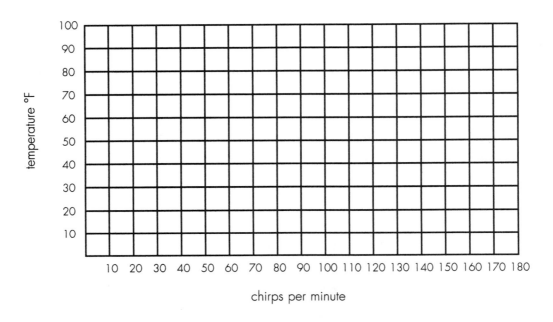

chirps per minute

From the graph, find the following:

1. the temperature when the crickets chirp 100 times/min.　　_____

2. the temperature when the crickets chirp 160 times/min.　　_____

3. the temperature when the crickets chirp 60 times/min.　　_____

4. the number of chirps/min. when the temperature is 60°F　　_____

5. the number of chirps/min. when the temperature is 70°F　　_____

6. the number of chirps/min. when the temperature is 50°F　　_____

Write an equation that shows the relationship between the chirps per minute and the temperature.

Decibels

Student Activity: "From Rustling Leaves to Jet Planes"—Students use given data to answer questions concerning decibel levels.

Background

Loudness is measured in decibels. A *decibel* (db) is the slightest difference in sound that can be detected by the human ear. Sound intensities increase by powers of 10. A 50-db sound is 10 times as loud as a 40-db sound, 100 times louder than a 30-db sound, and 1,000 times louder than a 20-db sound. Divide the difference of decibels by 10. The result is the power of 10.

$(50 - 40)/10 = 1$ → $10^1 = 10$ times as loud

$(50 - 30)/10 = 2$ → $10^2 = 100$ times as loud

$(50 - 20)/10 = 3$ → $10^3 = 1,000$ times as loud

SOUND	DECIBELS
jet plane	140
rock concert	120
thunder	90
subway	80
busy traffic	70
conversation	60
quiet car	50
light traffic	40
quiet conversation	30
whisper	20
rustle of leaves	10

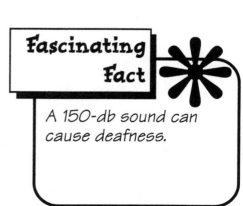

The intensity decreases as you move away from the source. As you double the distance, the decibel reading drops 6 db. If a sound at 4 ft. has a decibel reading of 25 db, at 8 ft., the decibel reading will be 19 db.

Fascinating Fact

A 150-db sound can cause deafness.

From Rustling Leaves to Jet Planes

D

Loudness is measured in decibels. A *decibel* (db) is the slightest difference in sound that can be detected by the human ear. Sound intensities measured in decibels increase by powers of 10. A 50-db sound is 10 times louder than a 40-db sound, 100 times louder than a 30-db sound, and 1,000 times louder than a 20-db sound.

Example: $(50 - 40)/10 = 1 \rightarrow 10^1 = 10$ times as loud

$(50 - 30)/10 = 2 \rightarrow 10^2 = 100$ times as loud

$(50 - 20)/10 = 3 \rightarrow 10^3 = 1,000$ times as loud

A sound 10,000 or 10^4 times louder will have a reading 40 db higher.

A sound $1/100$ or $1/10^2$ as loud will have a reading 20 db lower.

SOUND	DECIBELS
jet plane	140
rock concert	120
thunder	90
subway	80
busy traffic	70
conversation	60
quiet car	50
light traffic	40
quiet conversation	30
whisper	20
rustle of leaves	10

Fill in the blanks; then replace each letter in the expression with its corresponding answer to determine the decibel level that can cause deafness.

1. A jet plane is A_____ times louder than a rock concert.

2. A jackhammer is a thousand times louder than a subway. It has a decibel level of B_____.

3. If a clarinet has an intensity of 85 db, an object $1/100$ as loud would have an intensity of C_____ db.

4. A whisper is D_____ as loud as light traffic, whereas busy traffic is E_____ louder than light traffic.

5. Thunder is F_____ as loud as a subway.

6. Intensity decreases by 6 db when the distance is doubled. A sound of 31 db at 4 ft. would be G_____ at 8 ft.

Fascinating Fact

Deafness can occur at $(B - A) \times E \times D + (C - G) + F$ decibels, which is _____ decibels.

Diamonds

Student Activity: "Diamonds Are Forever"—Students use data from the Mohs Hardness Scale for rocks and minerals to draw conclusions.

Background

Friedrich Mohs, an Austrian geologist, devised a hardness scale for 10 minerals and gems. Diamond, the hardest known natural mineral, was given the value of 10. Lesser values were given to nine other minerals (see the chart). Relative hardness of an unknown substance can be determined by whether the substance can scratch, or be scratched by, one of the known minerals. For instance, quartz can scratch fluorite but cannot be scratched by fluorite.

Hardness of a substance has to do with its internal structure and not chemical makeup. A diamond and graphite both contain carbon, but a diamond has a hardness of 10 and graphite has a hardness between 1 and 2.

Mohs Hardness Scale	
Diamond	10
Corundum	9
Topaz	8
Quartz	7
Orthoclase	6
Apatite	5
Fluorite	4
Calcite	3
Gypsum	2
Talc	1

Hardness of Common Items

Steel file	6.5
Glass	5.5 – 6
Copper penny	3
Fingernail	2.5

D Diamonds Are Forever

Friedrich Mohs, an Austrian geologist, devised a hardness scale for 10 minerals and gems. Diamond, the hardest known natural mineral, was given the value of 10. Lesser values were given to nine other minerals. A mineral with a greater hardness number can scratch a mineral of lesser hardness.

Mohs Hardness Scale	
Diamond	10
Corundum	9
Topaz	8
Quartz	7
Orthoclase	6
Apatite	5
Fluorite	4
Calcite	3
Gypsum	2
Talc	1

Use the Mohs Hardness Scale to answer the following questions.

1. What mineral can scratch topaz but not a diamond? _____

2. What hardness number(s) would be given to a mineral that could scratch apatite but not corundum? _____

3. A copper penny will scratch a piece of gypsum but not fluorite. What hardness number would it be given? _____

4. A steel file will mark a piece of orthoclase but not a piece of quartz. What number would be given to a steel file? _____

5. A geologist used her fingernail to scratch a mineral on the list. If a fingernail has a hardness number of 2.5, what mineral was the geologist studying? _____

6. List the mineral samples A through D in descending hardness order given the following facts:_____

    Sample A scratches fluorite and calcite.
    Sample B scratches quartz.
    Sample C scratches gypsum but not fluorite.
    Sample D scratches quartz but not Sample B.

7. Jamie had a glass gem and a diamond but did not know which was which. How could Jamie determine which "gem" was the diamond? _____

Equation Exercises

Math Topics: Linear equations, absolute value equations, quadratic equations, plotting points

Student Activity: "Equation Exercise Task Card"—The teacher leads the class in a series of "exercises" based on the graphs of equation families. Students working in small groups record resting and active heart rates, plot the data, and discuss the results.

Students record their resting heart rates using the pulse spot on the wrist or just under the jawline. They count the beats for 10 seconds and multiply by 6 for a minute heart rate. Have students stand and use their arms to show the graphs of equations such as $y = x$, $y = -x$, $y = x + 1$, $y = x - 1$, $y = |x|$, $y = -|x|$, $y = x^2$, $y = -x^2$, etc. Do the "graphs" quickly for several minutes. Have students take their active heart rate. Groups of students should plot their data (resting, active) and draw the line of best fit. Make certain students know ahead of time that the activity includes aerobic activity.

Background

Equation families have similar graphs. For instance, all equations of the form $y = mx$ with $m > 0$ are lines passing through the origin rising left to right. When $m < 0$, the line falls left to right. Equations of the form $y = mx + b$ have a vertical shift equal to b. Equations of the form $y = ax^2$ are parabolas opening up when $a > 0$, and opening down when $a < 0$. All absolute value equations are V-shaped opening up or down depending on the coefficient of the absolute value quantity.

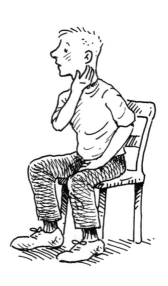

Equation Exercise Task Card

E

Tasks:

Use arms to model graphs of equation families.
Use an aerobic activity to collect data, plot points, and describe relationships.

Materials: Clock with a second hand

Activity:

Record your resting heart rate. Use the pulse point on the wrist or just under the jawbone. Count the beats for 10 seconds and multiply by 6 for the minute heart rate. Stand and use your arms to model the graphs of the equations given by the teacher or designated student. The activity should last a few minutes. Record the active heart rates. Plot the points (resting, active) for your group on a coordinate grid. Discuss the relationship.

Resting heart rate: _____ for 10 seconds; _____ for 60 seconds.

Graph the data points for your group.

Active

Resting

Discuss the relationship of resting heart rate to active heart rate. Write your conclusions.

Falling Objects

F

Math Topics: Evaluating an expression, rewriting an equation, square roots, negative integers

Student Activity: "B-52 Fallout Task Card"—In small groups or partners, students use a given formula to compute the time needed for falling objects to reach the ground.

Background

The force of gravity causes an object to fall 16 feet per second squared. In other words, the object accelerates by 16 ft./sec. for every second. An object's height is given by the formula $H_{final} = -16t^2 + H_{original}$

Example 1: An object dropped from 16 ft. will hit the ground in 1 second since
$$H_{final} = -16t^2 + H_{original} \rightarrow 0 = -16 (1)^2 + 16 \rightarrow 0 = -16 + 16$$

Example 2: An object dropped from 100 ft. will be 36 ft. from the ground after 2 seconds since
$$H_{final} = -16t^2 + H_{original} \rightarrow 36 = -16 (2)^2 + 100 \rightarrow 36 = -64 + 100$$

Fascinating Fact

Sound travels at 1,100 ft. per second. It would take one second for someone to hear an object 1,100 feet away. However, an object dropped from 1,100 ft. would take approximately 8 seconds to reach the ground since $0 = -16t^2 + 1,100 \rightarrow 16t^2 = 1,100$ $t^2 = 1,100/16 \rightarrow t = 8.3$ sec.

B-52 Fallout Task Card

Task:
Calculate the time it takes a falling object to hit the ground.

Materials:
Calculator with square root key

Activity:
On July 28, 1945, a B-52 bomber crashed into the seventy-eighth floor of the Empire State Building in New York City. If the floors are approximately 12.5 feet apart, how long did it take for the fallout of the plane crash to hit the ground below (height = 0)?

Information:
The height (H_{final}) of a falling object depends on the pull of gravity, the amount of time, and the original height ($H_{original}$) of the object. Gravity on earth causes an object to fall 16 ft. per second squared. Use the equation: $H_{final} = -16t^2 + H_{original}$. Substitute for H_{final} and $H_{original}$ and solve for t.

How much time did pedestrians have to get away after hearing the crash? _____

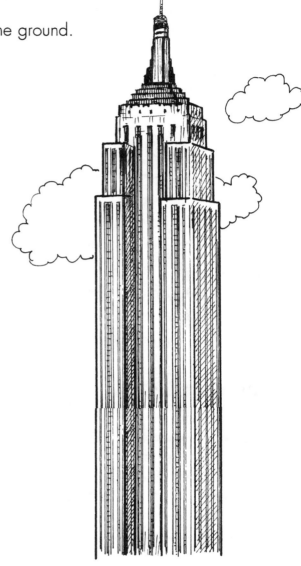

Fascinating Fact

Sound travels at 1,100 ft. per second. It would take one second for someone to hear an object 1,100 feet away. However, an object dropped from 1,100 ft. would take approximately 8 seconds to reach the ground since $0 = -16t^2 + 1,100$ ➜ $16t^2 = 1,100$ $t^2 = 1,100/16$ ➜ $t = 8.3$ sec.

Giants

Student Activity: "Giants: Fact or Fiction"—Students apply area and volume relationships of similar objects to data from the *Guinness Book of World Records* and *Gulliver's Travels* to discover some surprising facts concerning giants.

Background

Geometric shapes are similar when corresponding angles are congruent and corresponding sides are proportional. If the dimensions of a figure are doubled, the area is quadrupled. Since A = L x W, the resulting area would equal 2L x 2W or 4(L x W). If the dimensions of a figure are doubled, the volume is increased by a factor of 8. Since V = L x W x H, the resulting volume would be 2L x 2W x 2H or 8(L x W x H).

According to the *Guinness Book of World Records*, on June 27, 1940, Robert Wadlow was 8 feet 11.1 inches tall at the age of 22. He was roughly 1.5 times the height of the average male. His weight would be proportional to his volume (V = lwh). Therefore, his weight would be approximately 1.5^3 times that of an average male or 3.375 times as much, since $V_{wadlow} = 1.5l$ x 1.5w x 1.5h.

His bones (cross-section area) would have to support 1.5 times as much weight as an average male. Robert Wadlow had to wear leg braces to help support his weight. After an accident involving his brace, cellulitis developed and he died 18 days after his height was measured.

In *Gulliver's Travels* by Jonathan Swift, Gulliver visits the land of Brobdingnag where the inhabitants are 12 times as tall as Gulliver. Their feet would have an area 12^2 or 144 times as large supporting their weight, which would be 12^3 or 1,728 times as much as Gulliver's weight. The bones in their bodies would have to support 12 times as much weight as an average human.

Fascinating Fact

Consider the implications for the inhabitants of Lilliput who were 1/12 the height of Gulliver. The Lilliputians would weigh $1/12^3$ as much or 1/1,728 of Gulliver's weight. If you eat 2,000 calories, a Lilliputian would eat about about 1.2 calories.

Giants: Fact or Fiction

Fact: Robert Wadlow was a true giant. At age 22 he measured 8 feet 11.1 inches according to the *Guinness Book of World Records*. He was approximately 1.5 times as tall as a six-foot man.

1. If his feet were 1.5 times as long and 1.5 times as wide as those of a six-foot man, his feet would cover _____ times as much area (a = lw) as a six-foot man's foot.

2. Assuming food intake is proportional to volume (V = lwh), how many calories might he need if an average man consumes 3,000 calories? _____

Fiction: In *Gulliver's Travels* by Jonathan Swift, Gulliver travels to the land of Brobdingnag; the inhabitants there are similar to Gulliver but are 12 times as tall.

3. A Brobdingnagian's foot would cover _____ times as much area as Gulliver's foot.

4. If Gulliver uses two square yards of fabric to make a coat, how many yards would it take to make a Brobdingnagian's coat? _____

5. Weight is proportional to volume, so if you weigh 120 lbs., how much would a Brobdingnagian weigh? _____

6. If you consume 2,000 calories, how many calories would a Brobdingnagian consume? _____

7. Gulliver also visited Lilliput where the Lilliputians are $\frac{1}{12}$ the height of Gulliver. How much fabric would be needed to make a Lilliputian's coat? _____

8. If Gulliver consumes 3,000 calories, how many calories would a Lilliputian consume? (Assume calorie intake is proportional to volume.) _____

9. Could giants like the Brobdingnagians actually exist? _____ Why or why not? _____

Gravity

Math Topics: Ratios, proportions

Student Activity: "Weighty Matters"—Students calculate weights of items on other planets given the planetary gravity factors.

Background

Gravity is the force that attracts two bodies of matter towards each other. Sir Isaac Newton, an English mathematician, described it in the seventeenth century. The force of the attraction increases in proportion to the mass of the two objects and decreases in proportion to the square of the distance between the centers of the two objects. The mass of the sun is approximately 2×10^{30} kg. The total mass of all the planets in the solar system is less than 3×10^{27} kg. The mass of the sun is approximately 700 times the total mass of the planets. The gravitational pull of the sun keeps the planets in orbit.

Weight is the result of gravity. Since the pull of gravity on the moon is one-sixth the pull of gravity on Earth, someone weighing 120 pounds would weigh 20 pounds on the moon. The planets vary in mass and size. Therefore, the pull of gravity on each planet is different.

Planetary Gravity Factor Chart
Mercury = 0.37
Venus = 0.91
Earth = 1.00
Mars = 0.38
Jupiter = 2.53
Saturn = 1.07
Uranus = 0.92
Neptune = 1.18
Pluto = 0.09

Weighty Matters

Gravity is the force that attracts two bodies of matter towards each other. Weight is the measure of gravity's pull on an object. The greater the gravitational pull, the heavier the object. Since the planets vary in mass and size, the pull of gravity on each planet is different. Weight will vary from planet to planet.

Planetary Gravity Factor Chart
Mercury = 0.37
Venus = 0.91
Earth = 1.00
Mars = 0.38
Jupiter = 2.53
Saturn = 1.07
Uranus = 0.92
Neptune = 1.18
Pluto = 0.09

You can use the gravity factors to determine weights on other planets. For example, if you weigh 100 lbs. on Earth, you would weigh 100 x 0.37 = 37 lbs. on Mercury.

Use the chart to answer the following questions.

1. An astronaut weighs 140 lbs. on Earth. How much would she weigh on Venus? _____

2. An African elephant can weigh 6 tons (12,000 lbs.). How much would it weigh on Jupiter? _____
   On Pluto? _____

3. A rock on Mars weighs 19 lbs. How much would it weigh on Earth? _____

4. A rover sent to Pluto weighs 18 lbs. on the planet. Would you be able to lift the rover on Earth? _____ Why or why not? _____

   _____

5. How much would a 46-lb. Uranus object weigh on Saturn? _____
   On Neptune? _____

6. The gravitational pull of the moon is one-sixth the pull of Earth. How much would 25 lbs. of moon rocks (weighed on the moon) weigh on Earth? _____

Heart Disease

Student Activity: "Heart Disease—Risky Business"—Students calculate ratios to determine which sets of data indicate increased risk of heart disease.

Background

Most of the research on coronary artery disease, the heart problem that leads to most heart attacks, has been done on men. However, heart attack is the leading cause of death in American women. Often women's symptoms are ignored or misdiagnosed.

The heart pumping blood through the body provides oxygen and nourishment to the body's cells. The heart is a muscle. It weighs less than three pounds and beats more than 104,000 times a day. There are three coronary arteries that carry the blood. Coronary artery disease may occur when cholesterol deposits within the arteries block the heart muscle's supply of blood. The result of coronary artery disease may be angina, chest pains, or heart attack. Angina is similar to a heart cramp.

There are many risk factors including age, being overweight, high blood pressure, high cholesterol, lack of exercise, race, gender, smoking, stress, and other disease conditions such as diabetes.

Body type and size can indicate an increased risk of heart disease. For women, the risk increases if the ratio of waist measurement (w) to hip measure (h) is greater than 0.8. For men, the risk increases if the ratio is greater than 1.0.

Women: $w/h > 0.8$ Men: $w/h > 1.0$

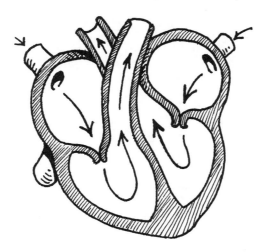

Heart Disease—
Risky Business

For women, the risk of heart disease increases when the ratio of the waist measurement to the hip measurement is greater than 0.8. For men, the risk increases when the ratio is greater than 1.0.

Women: w/h > 0.8   Men: w/h > 1.0

Compute the ratios **to the nearest hundredth**. Place the letter of the correct answer above the problem number below to discover what heart problem leads to most heart attacks. Circle the names of the people with an increased risk of heart disease.

1. Mr. Thomas: waist measure is 32"; hip measure is 34". _____

2. Miss Wilson: waist measure is 29"; hip measure is 34". _____

3. Mrs. Shehan: waist measure is 27"; hip measure is 36". _____

4. Mr. Talley: waist measure is 38"; hip measure is 37". _____

5. Mrs. Amera: waist measure is 30"; hip measure is 36". _____

6. Mr. Leary: waist measure is 34"; hip measure is 37". _____

7. Mrs. Garcia: waist measure is 28"; hip measure is 36". _____

8. Miss Zostel: waist measure is 35"; hip measure is 36". _____

9. Mr. Freeman: waist measure is 37"; hip measure is 35". _____

10. Mrs. Hoile: waist measure is 24"; hip measure is 34". _____

11. Mr. Mosier: waist measure is 40"; hip measure is 42". _____

ANSWERS:
A = 0.78
E = 0.92
I = 0.83
O = 1.06
U = 1.01
Y = 0.97
B = 0.70
C = 0.95
D = 1.03
G = 0.87
H = 0.99
K = 0.72
L = 1.05
M = 0.89
N = 0.75
R = 0.85
S = 0.94
T = 0.71

‾‾ ‾ ‾ ‾ ‾ ‾ ‾ ‾   ‾ ‾ ‾‾ ‾ ‾ ‾
11 9 2 9 3 7 2 8   7 2 10 6 2 8

‾ ‾ ‾ ‾ ‾ ‾ ‾
4 5 1 6 7 1 6

Interest

Student Activity: "Interesting Investments"—Students compare simple interest and compound interest.

Background

Interest is the amount of money paid to use someone's money. For instance, a bank pays interest on savings accounts so that it can use the deposited money to make loans for which it charges interest. The amount of money deposited or borrowed is called the *principal*.

Simple interest is interest paid on the principal only. Usually, it is calculated at the end of the stated time period for the account or the loan. The amount of interest equals the amount of the account or loan, called the principal, times the interest rate times the time in years. I = PRT

Ex.: $1,000 at 8% for three yrs. ➜ I = 1,000 × 0.08 × 3 ➜ I = $240
   The total amount (A) equals principal plus interest or $1,240.

Compound interest is interest paid on the principal and interest. It is calculated at various times during the term of the account or the loan.

Ex.: $1,000 at 8% for three yrs. compounded annually ➜ calculated each year
   1st year: $1,000 × 0.08 × 1 = $80 added to $ 1,000 = $1,080
   2nd year: $1,080 × 0.08 × 1 = $86.40 added to $1,080 = $1,166.40
   3rd year: $1,166.40 × 0.08 × 1 = $93.31 added to $1,166.40 = $1,259.71

The formula for the total amount when interest is compounded annually is
   $A = P(1 + R)^T$ $A = 1,000(1 + 0.08)^3$ ➜ A = $1,259.71

The formula for the amount when interest is compounded continuously is
   $A = P \times e^{RT}$ $A = 1,000 \times e^{0.08 \times 3}$ ➜ A = $1,271.25

Note: The number e is approximately equal to 2.718 if a calculator with an e^x key is not available.

Rule of 72: The rule of 72 states that when interest is compounded, the principal will be doubled when the product of the rate and the years equals 72.

   8% rate will double in 9 years   6% rate will double in 12 years

Name _____

Interesting Investments

Interest is the amount of money paid to use someone's money. The amount of money deposited or borrowed is the *principal*.

Simple interest is paid on the principal only. The interest equals principal times rate (%) times time (in years). I = PRT. The total amount (A) is principal plus interest.

Compound interest is interest paid on the principal and interest. The formula for the total amount when interest is *compounded annually* is $A = P(1 + R)^T$ The formula for the amount when interest is *compounded continuously* is $A = P \times e^{RT}$ Use 2.718 for *e* if a calculator with an e^x key is not available.

1. Amy invests $500 at 7% simple interest for 2 years. How much money will she have after 2 years?

2. If Amy invested $500 at 7% compounded annually, how much would she have after 2 years?

3. Shasha invests $200 at 5%. After 6 months, she puts all the money in a CD earning 7% for 12 months. How much money will she have?

4. Mr. Max cannot decide which plan would be the best, so he calculates simple, compounded annually, and compounded continuously interest for $2,000 at 8% for 3 years. Which is the best plan? How much more is earned over the other two plans?

Rule of 72 for compound interest:

When the rate percentage times the time equals 72, the amount is about doubled.

$1,000 at 8% will double in approximately 9 years. (8 x 9 = 72)

5. Tony has $500. How long will it take to have $1,000 if he invests at 6% compounded interest?

6. Al invests $2,000 in an IRA at age 24 at 6%. About how much will be in the account at age 60? Hint: How many times will it double?

Japanese Magic Circle

Math Topics: Consecutive integers, patterns

Student Activity: "Japanese Magic Circle Task Card"—In small groups or partners, students develop a strategy to determine the sum of the first 51 whole numbers using a magic circle.

Background

The Magic Circle shown was created by Seki Kowa, a seventeenth-century Japanese mathematician. In a Magic Circle, all diameters have the same sum. The sum of the magic circle is the product of the number of diameters and the sum of the numbers along one diameter (not including the center number), plus the center number.

Additional activities could include magic squares and magic triangles. A magic square is a square array of numbers such that each row, column, and diagonal has the same sum—the magic sum. The sum of all numbers in the magic square is the magic sum times the number of rows. A magic triangle has the same sum (magic sum) for the numbers along each edge. The sum of all numbers in a magic triangle is the magic sum times three minus the sum of the vertices.

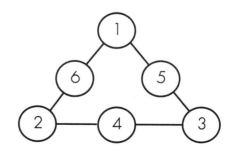

8	1	6
3	5	7
4	9	2

Magic Square
Sum = 15
Total = 45

Magic Triangle
Sum = 9
Total = 21

Japanese Magic Circle
Task Card

Task:

Determine the sum of the first 51 positive integers.

Materials:

The "magic circle" shown below.

Activity:

The "magic circle" shown below was developed by the Japanese mathematician Seki Kowa in the seventeenth century. Use it to determine the sum of the first 51 positive integers. Explain your method.

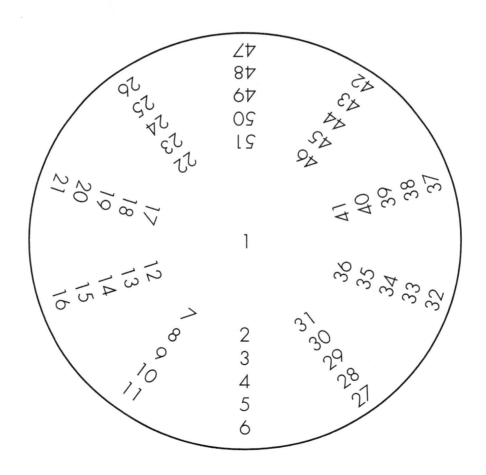

Kepler's Third Law

Math Topics: Negative exponents, constants, evaluating expressions, Calculator Use: Numbers in scientific notation

Student Activity: "Space Shuttle Task Card"—In small groups or partners, students use a given formula from Kepler's Third Law and data concerning satellites, the earth, and the space shuttle to compute the period of the space shuttle as it orbits the earth.

Background

Kepler's Laws explain planetary motion. They also explain the motion of the moon and satellites around the earth. The Third Law relates the period, P (time it takes for one revolution), to the distance, D, the satellite is from the center of the planet. The relationship is $P^2 = cD^3$ where c is the constant dependent on the size of the planet with the satellite orbiting. A *geosynchronous satellite* revolves the earth once every 24 hours. Geosynchronous satellites appear to stay in the same place above the earth. Knowing the distance of a geosynchronous satellite allows one to determine the constant c for earth. The value for c can then be used to determine the period for any object orbiting the earth at a given distance from the center of the earth.

Web site: http://physics.bu.edu/~gross/CGS/keplaw.html

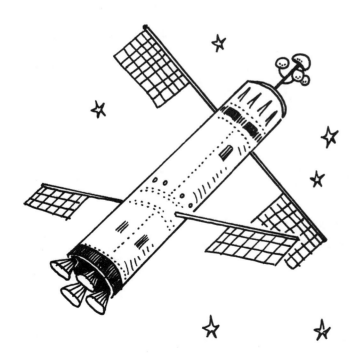

Fascinating Fact

Communication satellites are geosynchronous satellites, so a TV satellite dish can be pointed to one place in the sky and receive the signal from the satellite.

Space Shuttle Task Card

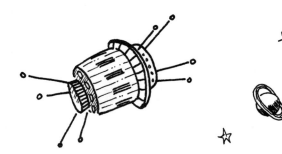

Task:
Determine the period of the space shuttle.

Materials:
Calculator

Activity:
The period, P, of a satellite (time for one revolution) is related to its distance, D, from the center of the earth by the expression: $P^2 = cD^3$ where c is a constant. This expression is known as Kepler's Third Law. A geosynchronous satellite has a period of 24 hours, so it stays at the same place above the earth. All geosynchronous satellites are located 44,250 km from the center of the earth.

1. Use the formula $p^2 = cD^3$ and the information for geosynchronous satellites to find the value for c. _____

2. The space shuttle orbits 250 km above the surface of the earth, which has a radius of 6300 km. Determine the distance the space shuttle is from the center of the earth.

   _____

3. Use the answers from #1 and #2 in the formula $p^2 = cD^3$ to determine the period of the space shuttle. _____

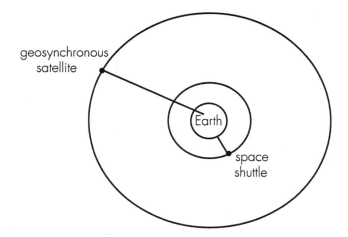

geosynchronous satellite

Earth

space shuttle

Fascinating Fact | *All communication satellites are geosynchronous satellites.*

Konigsberg's Bridges

Math Topics: Data analysis, networks, topology

Student Activity: "Konigsberg's Bridges Task Card"—In small groups or partners, students explore networks, figures containing vertices connected by arcs, to determine which networks can be drawn with one continuous stroke.

Background

The seven bridges of Konigsberg, Germany, suggest the problem of whether it is possible to cross each of the bridges exactly once. A diagram of the bridges can be represented by a network. The four regions of the city are shown as vertices connected by arcs (bridges).

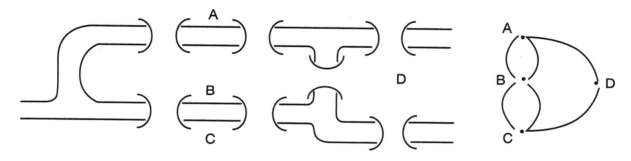

The degree of a vertex is the number of arcs that meet at the vertex. In the Konigsberg Bridge network all four vertices have odd degrees (A, C, D–3; B–5). In the 1700s, Leonhard Euler showed that a network can be drawn in one continuous stroke if it contains no more than two vertices of odd degree. Therefore, the Konigsberg Bridge network cannot be drawn in one continuous stroke.

A vertex of odd-degree must be the start or end of a continuous path. If an odd-degree vertex is not the start or end of a path, then when the path comes to the vertex it must exit the vertex (even number of arcs used) which leaves an arc unused. When that arc is used, the path is stuck at the vertex with "no way out." A path only has one start and one end, so the number of odd-degree vertices cannot exceed two.

The continuous path that connects the arcs of a network is called an *Euler* (Oy lur) *path*.

Konigsberg's Bridges
Task Card

Task:

Determine a rule for when a network can be drawn in one continuous stroke.

Background:

The seven bridges of Konigsberg, Germany, suggest the problem of whether it is possible to cross each of the bridges exactly once. A diagram of the bridges can be represented by a network. The four regions of the city are shown as vertices connected by arcs (bridges).

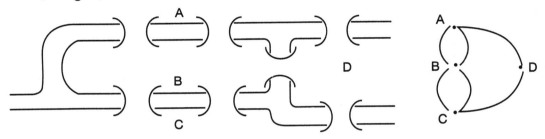

The degree of a vertex in a network is the number of arcs that meet at the vertex. In the Konigsberg Bridge network all four vertices have odd degrees (A, C, D–3; B–5).

Activity:

Copy the bridge network. Can you trace the network in one continuous stroke without retracing any arc? _____

Copy each network. Determine which ones can be traced in one continuous stroke without retracing any arc. Complete the table. What do the networks that cannot be traced have in common? _____

1.    2.

3. 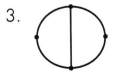   4.

5. 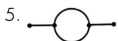   6.

7.   8.

Network	# even vertices	# odd vertices	Is it traceable?
1	3	2	yes
2	1	4	
3			
4			
5			
6			
7			yes
8			

Ladders

Student Activity: "Reaching New Heights"—Students solve problems involving ladders using the Pythagorean formula and explore the concept of the sine function.

Background

The foot of a ladder should be placed at least one fourth of the length of the ladder from a wall. For example, the end of a 12-foot ladder should be at least three feet from the wall. If a ladder is too steep, it will tip backwards when the weight of a person causes the center of gravity to shift behind the ladder.

The Pythagorean formula $c^2 = a^2 + b^2$ can be used to find the maximum height a ladder will reach. The ladder, wall, and the ground form a right triangle. The length (L) of the ladder is the hypotenuse, the distance (D) from the wall is one leg, and the height (H) the ladder reaches is the other leg. For ladders, $L^2 = D^2 + H^2$. The greatest height is reached when the foot of the ladder is as close to the wall as possible ($D = \frac{1}{4}L$).
$L^2 = D^2 + H^2 \rightarrow L^2 = (\frac{1}{4}L)^2 + H^2 \rightarrow L^2 - (\frac{1}{4}L)^2 = H^2 \rightarrow \frac{15}{16}L^2 = H^2 \rightarrow \frac{\sqrt{15}}{4}L = H$.

In a right triangle, the ratio of the leg opposite an angle to the hypotenuse is called the *sine* of the angle. The ratio of the maximum height of the ladder to the length of the ladder is $\frac{\sqrt{15}}{4}L$ to L or $\frac{\sqrt{15}}{4}$ which equals 0.968. The sine of a 75° angle equals 0.966. The maximum angle a ladder should make with the ground is 75°.

Reaching New Heights

For safety reasons, the foot of a ladder should be at least one fourth of the length of the ladder from a wall. When leaning against a wall, a ladder is the hypotenuse of a right triangle having the wall and the ground as the two legs.

The Pythagorean formula $c^2 = a^2 + b^2$ becomes $L^2 = H^2 + D^2$, when L is the length of the ladder, H is the height it reaches, and D is the distance from the wall.

Use the information given to answer the following questions.

1. What is the least distance a 12-foot ladder should be placed from the base of a wall?

   _____

2. Will the ladder in problem #1 reach a 10-foot high window? Why or why not? _____

3. A flower garden extends 5 feet from the base of a wall. The foot of a ladder must be placed on the sidewalk border. How far up the wall will a 15-foot ladder reach?

   _____

4. To reach a ledge 20 feet above the ground, what is the shortest ladder that can safely be used?_____

5. Make a scale drawing of a 12-foot ladder leaning up against a wall with the foot of the ladder the minimum safe distance from the wall. Let 1 cm = 2 ft. Measure the angle the ladder makes with the ground. _____

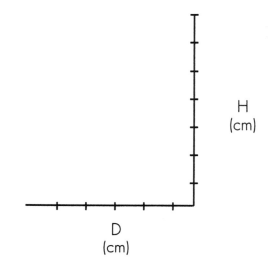

H
(cm)

D
(cm)

Lightning and Thunder

Math Topics: Scientific notation, units conversions

Student Activity: "Storm Tracker Task Card"—In small groups or partners, students complete a series of units conversions pertaining to the speed of light and the speed of sound that lead to a method of determining the distance of an electrical storm.

Background

Light travels at 186,000 mi./sec. or approximately 982,000,000 ft./sec. (9.82×10^8 ft./sec.). Sound travels at 1,100 ft./sec. In metric units, light travels at 300,000,000 m/sec. or 3×10^8 m/sec. and sound travels at 330 m/sec. or 3.3×10^2 m/sec. A lightning strike (light) is seen almost immediately. The sound of the resulting thunder can take several seconds to be heard depending on the distance of the lightning strike.

The sound of thunder travels 1,100 ft./sec. There are 5,280 ft. in a mile, so the sound travels approximately 1 mile in 5 seconds. A count of 5 seconds between the lightning strike and the clap of thunder means the lightning strike and storm are one mile away. A count of 10 seconds means the storm is 2 miles away, etc.

In metric measures, the sound of thunder travels 330 m/sec. or 0.33 km/sec. Therefore, the sound travels 1 km in 3 seconds. A count of 3 seconds between the lightning strike and the sound of the thunder means the storm is 1 km away. A count of 6 seconds means the storm is 2 km away, etc.

(L) Storm Tracker Task Card

Task:
Discover a method for determining the distance to an electrical storm.

Materials:
Calculator

Activity:
Light travels at 186,000 mi./sec.

1. Convert the speed of light to ft./sec. _____

2. Round to the nearest million ft./sec. _____

3. Write the rounded speed in scientific notation. _____

Lightning travels at the speed of light, so the lightning is seen almost instantly.

Sound travels at 1,100 ft./sec. There are 5,280 feet in one mile.

4. Compute the time it takes for a sound to travel 1 mile. _____

5. Round the time to the nearest second. _____

6. A _____-second count indicates a distance of one mile.

Thunder travels at the speed of sound.

In metric units, sound travels at 330 m/sec.

7. Convert the speed of sound to km/sec. _____

8. Compute the time it takes for a sound to travel 1 kilometer. _____

9. A _____ second count indicates a distance of one kilometer.

10. Use the information to describe a method for determining the distance of a storm.

Math Magic

Math Topics: Place value, number sense

Student Activity: "Magic Card Task Card"—In small groups or partners, students will be able to predict the top card in a deck of cards after performing a series of steps. The "trick" can be explained using number sense concepts.

Background

A preselected card is placed in a deck of cards at the eighteenth position from the top. When the deck of cards is "cut in half" usually 23 to 29 cards will remain in the top "half." After the cards are counted out from top to bottom, the preselected card is eighteenth from the bottom of the deck. To have the preselected card on top, cards must be removed from the top of the deck. The number of cards to remove is equal to the sum of the digits of the number of cards in the deck. Eighteen cards will remain, so the top card is the preselected card.

Eighteen cards remain because when the sum of the digits $(T + U)$ of a two-digit number $(10T + U)$ is subtracted from the number, the difference is a multiple of 9: $(10T + U) - (T + U) = 9T$. When the two-digit number is between 20 and 29, T equals 2 and 9T equals 18.

Magic Card Task Card

Task:

Perform a magic trick with a deck of cards and predict the top card after a series of steps.

Materials:

A standard deck of playing cards

Activity:

1. Choose a card to be the Magic Card. With the deck face down, secretly place the card so it is the eighteenth card from the top of the deck.

2. Have someone "cut the deck in half."

3. Discard the bottom half.

4. Have someone count out the number of cards remaining. Physically count out the cards so the top card is placed down first making it the bottom card.

5. Find the sum of the digits of the number of cards counted.

6. Have someone remove that number of cards from the top of the deck.

7. Predict the top card—it will be the preselected magic card.

Review the steps and try to explain how the magic card makes it to the top of the deck.

Medical Math

Math Topic: Conversion of measures

Student Activity: "Med-Math Shortcut"—Students approximate the number of kilograms given the weight in pounds using a method used by medical professionals.

Background

One pound equals 2.2 kilograms, so converting from kilograms to pounds is not difficult—multiply by 2.2. However, to convert from pounds to kilograms, one must divide by 2.2. It is not easy to divide by 2.2 mentally. Medical professionals, such as paramedics, often have to approximate a patient's weight in terms of kilograms to administer medical dosages that are prescribed per kilogram. Paramedics do not have the time to use a calculator. The following is a shortcut for approximating weights in kilograms.

Take half the weight in pounds and subtract one tenth of that result.

Example 1: Shortcut: 180-lb. man ➔ ½ of 180 = 90
  ¹⁄₁₀ of 90 = <u>- 9</u>
  81 kg

  Actual: Dividing 180 lbs. by 2.2 lbs./kg results in 81.8 kg

Example 2: Shortcut: 120-lb. woman ➔ ½ of 120 = 60
  ¹⁄₁₀ of 60 = <u>- 6</u>
  54 kg

  Actual: Dividing 120 lbs. by 2.2 lbs./kg results in 54.5 kg.

Using the shortcut results in a 1% error when the given weight is 220 lbs. The actual conversion is 100 kg. The shortcut conversion is 99 kg.

Med-Math Shortcut

Medical professionals, such as paramedics, often have to make quick numerical conversions without the use of a calculator. Medications are prescribed according to the weight of the patient in kilograms. To convert from pounds to kilograms, one divides by 2.2 lbs./kg. The result is not easily computed mentally.

A shortcut for a good approximation of kilograms when given weight in pounds is to take half of the weight in pounds and subtract one tenth of the result.

Example: 180-lb. man ➔   ½ of 180 = 90
                                          $^-$⅒ of 90 = - 9
                                                       ――――
                                                         81 kg

Use the shortcut to approximate the kilogram for each weight in pounds.

1. 120-lb. woman _____   2. 140-lb. man _____

3. 72-lb. child _____   4. 160-lb. man _____

5. ton (2,000 lbs.) _____   6. 10 lbs. _____

7. 500 lbs. _____   8. 40 lbs. _____

9. 80 lbs. _____   10. 340 lbs. _____

Extension:

220 lbs. equals 100 kg. The shortcut results in 220 lbs. = _____

What is the percent difference in the two results? _____

Miniature Golf

Math Topics: Plotting points, angles, reflections, using a ruler

Student Activity: "Puttering Around"—Students use points of reflection to determine the path of a miniature golf ball for a hole-in-one.

Background

Disregarding friction, the angle of incidence equals the angle of reflection when a ball banks off a side railing in miniature golf. The same is true for light or laser beams reflecting off a mirror or billiard balls banking off cushions.

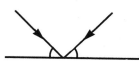

When a barrier or obstacle prevents a direct shot for a hole-in-one, banking a shot may result in a hole-in-one. Reflecting the hole over the axis determined by the rail gives an image of the hole, H_i. Aiming at the hole image will result in a hole-in-one. Notice that in the diagram, $\angle 1 = \angle 2$ (vertical angles are congruent), and $\angle 2 = \angle 3$ (angle of incidence equals angle of reflection). Therefore, $\angle 1 = \angle 3$. The ball will go into the hole.

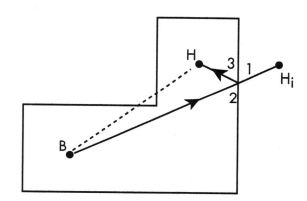

If the path to the hole image is blocked, a second reflection about another rail is needed. The resulting shot is a double bank shot.

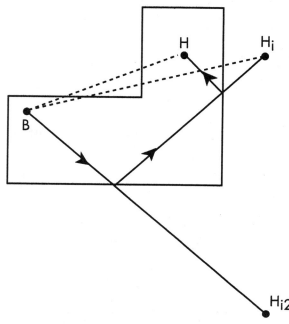

Puttering Around

When a barrier prevents a direct shot for a hole-in-one in miniature golf, banking a shot may result in a hole-in-one.

To determine the path of the ball:

1. Reflect the hole over the axis determined by the rail to obtain an image of the hole,

2. Draw the path from the ball to the hole image.

3. Aim at the point where the path intersects the rail.

4. The ball will bank into the hole—hole-in-one.

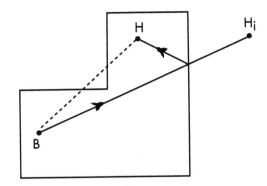

For each figure, reflect the hole (H₁) and draw the path for a hole-in-one.

1.

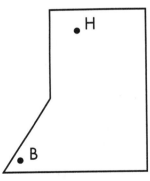

2.

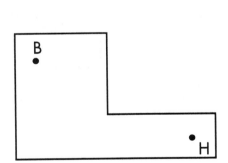

3.

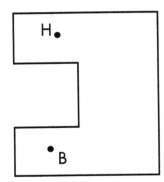

In figure 4, a corner prevents a hole-in-one in one bank shot. Reflect the hole image over another rail. Draw the line from the ball to your second reflected hole image. Draw the line from the intersection with the rail to the first hole image for a double bank shot.

4.

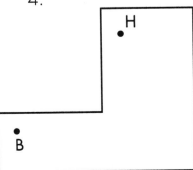

Nautilus Shell

Math Topics: Data collecting, metric measurement, golden ratio

Student Activity: "Chambered Nautilus: A Golden Shell"—

I. Students measure the length and width of chambers of a nautilus shell and compare the ratio of length to width to the golden ratio (1.618).

II. Students explore a personal golden ratio.

III. Students survey rectangle preferences to illustrate the eye-pleasing feature of golden rectangles.

Background

The spiral of a chambered nautilus shell will fit in repeated golden rectangles. A golden rectangle has length and width dimensions which form the ratio of 1.618.

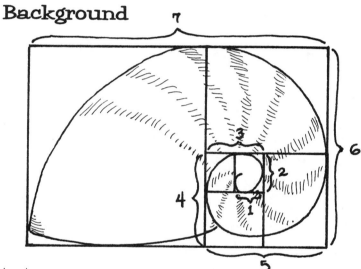

A personal golden rectangle is found when you curl your first finger: compare the distance from your knuckle to the first joint to the distance from the first to second joint.

Golden rectangles are considered the most pleasing to the eye and are commonly used in architecture and photography. An 8" x 5" photograph approximates the golden ratio—$\frac{8}{5}$ equals 1.6. The sides of a golden rectangle will follow the proportion:

$$\frac{L}{W} = \frac{L + W}{L}$$

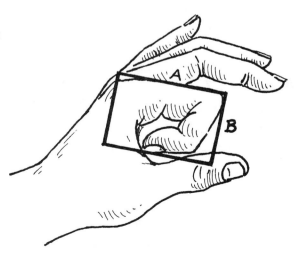

Chambered Nautilus: A Golden Shell

N

The *golden rectangle* is a rectangle which has a length-to-width ratio of 1.618. Golden ratios are found throughout nature, and golden rectangles are used in architecture.

Activity I:

1. Use a metric ruler.
2. Measure the length and width of each rectangle containing part of the spiral of the nautilus shell.
3. Calculate the ratios L/W.
4. How is the spiral of the nautilus shell related to the golden ratio.

	L (mm)	W (mm)	L/W
1.	(2)	(1)	
2.	(3)	(2)	
3.	(4)	(3)	
4.	(5)	(4)	
5.	(6)	(5)	
6.	(7)	(6)	

Activity II:

1. Curl your first finger. See the diagram.
2. Measure the distance from your knuckle to the first joint. (A)
3. Measure the distance from the first to the second joint. (B)
4. Calculate the ratio A/B. Compare this ratio to the golden ratio.

Activity III:

1. Draw rectangles 4 cm x 5 cm, 5 cm x 8 cm, and 8 cm x 10 cm.
2. Ask 15 members of your class which shape they prefer.
3. Record your results.
4. Compute the ratio of the length to the width for each rectangle.
5. Which "ratio" was preferred?

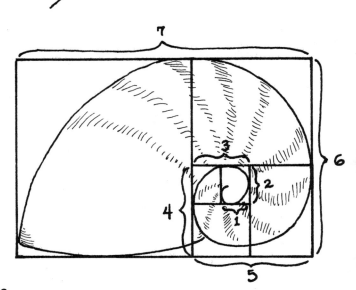

Nutrition Labels

Student Activity: "Nutrition Labels: You Are What You Read"—Students compute answers for nutrition problems using information given in nutrition labels.

Background

Nutrients are divided into five categories: carbohydrates, fats, proteins, vitamins, and minerals. Carbohydrates and proteins supply four calories per gram, and fat provides nine calories per gram.

Nutrition labels state the recommended serving size and number of servings in the container besides supplying nutritional information. The amounts of calories, total fat, saturated fat, cholesterol, sodium, carbohydrates, dietary fiber, sugars, and protein per serving are given. The Percent Daily Value quantities are based on a 2,000-calorie diet. Vitamin A, vitamin B, calcium, and iron percents are listed.

SOUP
Nutrition Facts:
Serving Size 1 cup (240 ml)
Servings about 2.5

Calories 70		from fat 20
Total Fat	2 g	3%
Sat. Fat	1 g	5%
Cholesterol	15 mg	5%
Sodium	980 mg	41%
Total Carb.	9 g	3%
Fiber	1 g	4%
Sugars	1 g	
Protein	3 g	
Vitamin A		6%
Vitamin C		0%
Calcium		2%
Iron		4%

STEW
Nutrition Facts:
Serving Size 1 cup (236 g)
Servings 3

Calories 230		from fat 120
Total Fat	14 g	22%
Sat. Fat	7 g	35%
Cholesterol	40 mg	13%
Sodium	950 mg	40%
Total Carb.	16 g	5%
Fiber	2 g	8%
Sugars	3 g	
Protein	11 g	
Vitamin A		6%
Vitamin C		0%
Calcium		2%
Iron		6%

Students could collect food labels and compare no-fat, low-fat, and regular food items. Often lower fat items have more calories due to increased sugar content.

Nutrition Labels: You Are What You Read

Nutrition labels state the recommended serving size and number of servings in the container besides supplying nutritional information per serving. The Percent Daily Value quantities are based on a 2,000-calorie diet.

SOUP
Nutrition Facts:
Serving Size 1 cup (240 ml)
Servings about 2.5

Calories 70	from fat 20	
Total Fat	2 g	3%
Sat. Fat	1 g	5%
Cholesterol	15 mg	5%
Sodium	980 mg	41%
Total Carb.	9 g	3%
Fiber	1 g	4%
Sugars	1 g	
Protein	3 g	
Vitamin A		6%
Vitamin C		0%
Calcium		2%
Iron		4%

STEW
Nutrition Facts:
Serving Size 1 cup (236 g)
Servings 3

Calories 230	from fat 120	
Total Fat	14 g	22%
Sat. Fat	7 g	35%
Cholesterol	40 mg	13%
Sodium	950 mg	40%
Total Carb.	16 g	5%
Fiber	2 g	8%
Sugars	3 g	
Protein	11 g	
Vitamin A		6%
Vitamin C		0%
Calcium		2%
Iron		6%

Use the two nutritional labels to answer the following questions.

1. What percent of the calories for the soup are from fat? _____

2. What percent of the calories for the stew are from fat? _____

3. Which item has less cholesterol? _____

   How many fewer mg/serving? _____

4. If someone is watching her sodium count, which item would be better? _____

5. If 3 g of fat is 5% of the Percent Daily Value, how much fat is "allowed" in a 2,000-calorie diet? _____

6. Which item has more protein? _____

   How many more grams/serving? _____

7. Which nutrient does neither item contain?_____

8. A diabetic counts carbs. A *carb* is 15 g of carbohydrates. How many *carbs* does a serving of each item contain?_____

Orbits

Math Topics: Nonstandard measures, conversion of measures

Student Activity: "Outdoor Orbits Task Card"—Ten students model the distances between the Sun and the planets of the solar system and convert the model to actual distances.

Background

The planets travel in elliptical orbits around the sun. The activity models a Grand Conjunction, when the planets are in a line. It should be noted that the model is a rough estimation of the distances between the planets. Additional students could be used to represent asteroids and moons.

The model results in the following distances:

		Cumulative Distance
1 step from Sun to Mercury	31,000,000 mi.	31,000,000 mi.
1 step from Mercury to Venus	31,000,000 mi.	62,000,000 mi.
1 step from Venus to Earth	31,000,000 mi.	93,000,000 mi.
1½ steps from Earth to Mars	46,500,000 mi.	139,500,000 mi.
10½ steps from Mars to Jupiter	325,500,000 mi.	465,000,000 mi.
12½ steps from Jupiter to Saturn	387,500,000 mi.	852,500,000 mi.
28 steps from Saturn to Uranus	868,000,000 mi.	1,720,500,000 mi.
31½ steps from Uranus to Neptune	976,500,000 mi.	2,697,000,000 mi.
27½ steps from Neptune to Pluto	852,500,000 mi.	3,549,500,000 mi.

Fascinating Fact

Since 1980 the orbits of Neptune and Pluto are such that Neptune is farther from the Sun than Pluto. In the year 2000, Pluto will again be the farthest planet. A mnemonic for remembering the order of the planets Mercury, Venus, Earth, Mars, Jupiter, Saturn, Uranus, Neptune, and Pluto (except from 1980 to 2000) is: **My Very Eager Mother Just Sewed Uncle Ned's Pants.**

Outdoor Orbits Task Card

Task:

Represent the planetary distances of the Solar System.

Materials:

Ten students each with a sign (Sun, Mercury, Venus, Earth, Mars, Jupiter, Saturn, Uranus, Neptune, and Pluto)

a large area outdoors

Activity:

One student holds the sign labeled Sun. The student holding the Mercury sign takes one step away from the Sun. The other students take their positions according to the list:

1 step from Sun to Mercury   _____

1 step from Mercury to Venus   _____

1 step from Venus to Earth   _____

1½ steps from Earth to Mars   _____

10½ steps from Mars to Jupiter   _____

12½ steps from Jupiter to Saturn   _____

28 steps from Saturn to Uranus   _____

31½ steps from Uranus to Neptune   _____

27½ steps from Neptune to Pluto   _____

If each step represents approximately 31 million miles, calculate the various distances.

Fascinating Fact

Between 1980 and 2000, the orbits of Neptune and Pluto are such that Neptune is farther away from the Sun than Pluto.

Ozone

Math Topics: Data analysis, percent decrease (increase)

Student Activity: "Ozone Destroyers—CFCs"—Students use given data concerning CFC (chlorofluorocarbons) use from 1976 to 1992 to compute percents of increase and decrease. Calculators are recommended.

Background

Ozone (O_3) is a naturally occurring gas. Ozone molecules contain three oxygen atoms, whereas the oxygen molecules we breathe (O_2) contain two oxygen atoms. The ozone that exists in the troposphere, the lower atmosphere, is considered "bad ozone" because it is the major component of smog. The ozone existing in the stratosphere, the upper atmosphere, is necessary for life on the surface of the earth since it protects the earth from the sun's ultraviolet radiation.

The ozone layer is between 12 and 30 miles above the earth's surface. The ozone layer completely absorbs UV-C radiation, which is lethal to humans, and 95-98% of UV-B radiation, which is known to cause skin cancer. Natural formation and destruction of ozone molecules results from the absorption of UV light. When UV light hits an oxygen molecule, the molecule is broken apart. The single atoms then combine with oxygen molecules to form ozone. The process continues:

$$O_2 + \text{UV light} = O + O$$
$$O + O_2 = O_3$$
$$O_3 + \text{UV light} = O_2 + O$$

Fascinating Fact

One CFC molecule can destroy 100,000 ozone molecules. Source: http://www.foe.org/ozone/intro.html

Destruction of the ozone is accelerated by certain synthetic chemicals including chlorofluorocarbons (CFCs). CFCs are used in refrigerant coolants, cleaning solvents, and aerosol propellants. CFC use in aerosol sprays dropped following a 1978 ban by the Environmental Protection Agency.

Students should be reminded that the percent of change (decrease or increase) is always the ratio of the amount of change to the original (or starting) quantity.

Ozone Destroyers—CFCs

Chlorofluorocarbons (CFCs) are synthetic chemicals used in refrigerant coolants, cleaning solvents, and aerosol propellants. CFCs are called "ozone destroyers" because UV radiation causes the molecules to break down and release chlorine, which destroys ozone faster than it can naturally be created.

CFC USE BY TYPE IN THE UNITED STATES

Year	Refrigeration	Blowing Agent	Aerosol Propellant	Other Uses
1976	154,720	111,948	432,275	51,617
1980	191,280	173,318	243,217	32,023
1984	211,439	223,148	218,834	41,041
1988	267,856	312,024	176,550	40,558
1992	195,178	169,745	22,866	14,852

Use the data to determine the following percents of increase or decrease.

1. Refrigeration from 1976 to 1988 _____ increase or decrease

2. Refrigeration from 1988 to 1992 _____ increase or decrease

3. Blowing Agent from 1976 to 1988 _____ increase or decrease

4. Blowing Agent from 1988 to 1992 _____ increase or decrease

5. Aerosol Propellant from 1976 to 1992 _____ increase or decrease

6. Other Uses from 1980 to 1984 _____ increase or decrease

7. Other Uses from 1988 to 1992 _____ increase or decrease

Piano

Student Activity: "Piano Task Card"—In small groups or partners, students explore the frequencies of notes that are octaves apart.

Background

The A below middle C (A-220) on a piano has a frequency of 220 vibrations per second. The frequency, F, of any note can be found by the formula: $F = 220 \left(2^{2/12} \right)$ where n is the number of keys above (positive) or below (negative) A-220. Think of the keyboard as a number line with A-220 at the origin.

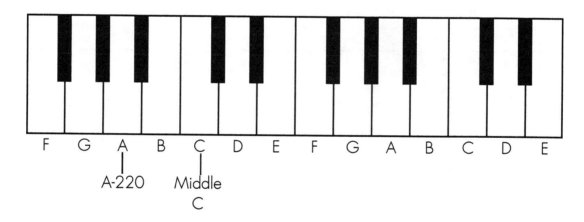

An *octave* is eight notes. The octave from A-220 to the next A contains the notes A, B, C, D, E, F, G, and A. However, starting at A there are 12 keys from one A to the next A. When n = 12, $2^{12/12} = 2^1 = 2$. The frequencies of octaves vary by a factor of 2.

Name _____

Piano Task Card

Task:
Determine and compare the frequencies of various piano notes.

Materials:
Calculator with a power key (y^x or $\wedge$)

Activity:
The A below middle C (A-220) on a piano has a frequency of 220 vibrations per second. The frequency, F, of any note can be found by the formula:

$$F = 220 \left(2^{n/12} \right)$$

where n is the number of keys above (positive) or below (negative) A-220. Think of the keyboard as a number line with A-220 at the origin.

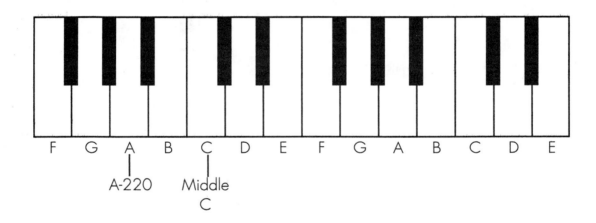

1. Find the frequency of the A above middle C. _____
   (Be certain to count the white and black keys.)

2. Find the frequency of middle C. _____
   (Hint: Use a calculator with a power key (y^x or $\wedge$).)

3. Find the frequency of high C. _____

4. How do the frequencies of notes an octave apart compare? _____

Potatoes

Math Topics: Rewriting a formula, distributive property, writing numbers in terms of thousands, evaluating an expression

Student Activity: "Hot Potato"—Students determine the time it takes to boil a potato at various altitudes. Calculators are recommended.

Background

At sea level and standard pressure, the boiling point of water is 212°F. At higher elevations, due to a decrease in pressure, water boils at a lower temperature.

A formula for the boiling point, B, of water at x thousand feet is

$$B = 212 - 1.85x$$

At lower temperatures, the time it takes to "boil a potato" increases.
The time, T, to boil a potato is

$$T = 40(237 - B)/(B - 162)$$

Substituting the value for B gives the formula

$$T = 40(25 + 1.85x)/(50 - 1.85x)$$

for the number of minutes to boil a potato at x thousand feet.

Fascinating Fact ✳

According to *The Potato Book* by Myrna Davis, one acre of land can produce 600 bushels of potatoes. The average is about 300 bushels.

Hot Potato

P

At sea level and standard pressure, the boiling point of water is 212°F. At higher elevations, the boiling point of water decreases. At a lower boiling point, it takes longer to boil a potato.

a formula for the boiling point of water at x thousand feet is

1. $B = 212 - 1.85x$

a formula for the time, T, in minutes to boil a potato is

2. $T = \dfrac{40\,(237 - B)}{B - 162}$

Substitute the value for B from formula 1 into formula 2 and simplify:

3. $T = \dfrac{40\,(237 - (\underline{\hspace{3cm}}))}{(\underline{\hspace{3cm}}) - 162}$

4. $T = \dfrac{40\,(\underline{\hspace{3cm}})}{(\underline{\hspace{3cm}})}$

Use formula 4 to determine the time, T, to the nearest minute to boil a potato:

Location	x thousand feet	minutes to nearest tenth
5. Albuquerque, New Mexico, at 4,945 ft.	4.945	
6. Boston, Massachusetts, at 21 ft.		
7. Flagstaff, Arizona, at 6,900 ft.		
8. Helena, Montana, at 4,155 ft.		
9. Las Vegas, Nevada, at 2,030 ft.		
10. Omaha, Nebraska, at 1,040 ft.		

11. Give two locations that have an elevation difference of approximately 1,000 feet:

_____ and _____.

What is the difference in time? _____

Quilt Design I

Math Topics: Modular arithmetic, reflections

Student Activity: "Mod-Math Quilt Task Card"—In small groups or partners, students complete a modular math addition table, replace the numbers with designs, and reflect the pattern to create a quilt design. Groups need metric rulers and paper.

Background

Quilt designs are often basic geometric shapes or patterns that are reflected about vertical and horizontal axes. Modular math systems have a finite set of elements and can be used to generate a quilt pattern.

Mod 5 uses the numbers 0, 1, 2, 3, and 4. If the sum or product of two numbers is greater than 4, the answer is the remainder when the result is divided by 5. For example, 2 + 4 = 1 mod 5 (2 + 4 = 6 and 6/5 has a remainder of 1) and 2 x 4 = 3 mod 5 (2 x 4 = 8 and 8/5 has a remainder of 3). Since the system has a finite number of elements, symbols can be used to create a pattern/art design.

The task card has students complete an addition and a multiplication table modular 5. Students choose one table as a basis for their quilt pattern. Students then replace each number with a design. The pattern is then reflected to the right, then down, and finally to the left. If the pattern is placed in the second quadrant, then the first reflection is about the positive y-axis. The resulting reflection is then reflected about the positive x-axis. The final reflection is either the fourth quadrant reflected about the negative y-axis or the second quadrant reflected about the negative x-axis. The results of the various reflections could be a topic for class discussion.

Students could use irregular grids.

+	0	1	2	3	4
0	0	1	2	3	4
1	1	2	3	4	0
2	2	3	4	0	1
3	3	4	0	1	2
4	4	0	1	2	3

x	0	1	2	3	4
0	0	0	0	0	0
1	0	1	2	3	4
2	0	2	4	1	3
3	0	3	1	4	2
4	0	4	3	2	1

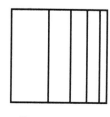

Converging

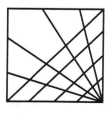

Kaleidoscopic

Name _____

Mod-Math Quilt
Task Card

Task: Design a quilt based on a mod-math table.

Materials: Metric ruler, paper

Background:
Modular arithmetic is a system using a finite set of numbers and an operation. Mod 5 uses the numbers 0, 1, 2, 3, and 4. If the sum or product of two numbers is greater than 4, the answer is the remainder after dividing by 5.

Examples: $2 + 4 = 1$ mod 5 ($2 + 4 = 6$ and $\frac{6}{5}$ has a remainder of 1)
 $2 \times 4 = 3$ mod 5 ($2 \times 4 = 8$ and $\frac{8}{5}$ has a remainder of 3)

Activity:
1. Complete the addition and multiplication tables mod 5.
2. Draw a symbol for each number.
3. Replace the numbers in the tables with the symbols.
4. Draw a 20 cm x 20 cm square divided into 100 2 cm x 2 cm cells.
5. Reproduce your symbol table in the upper left quadrant.
6. Reflect your design into the upper right quadrant. Reflect your design into the lower right quadrant. Lastly, reflect your design into the lower left quadrant.

1.
+	0	1	2	3	4
0					
1					
2					
3					
4					

x	0	1	2	3	4
0					
1					
2					
3					
4					

2.
0 = ☐

1 = ☐

2 = ☐

3 = ☐

4 = ☐

3. +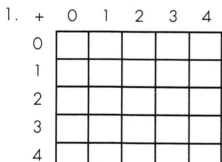

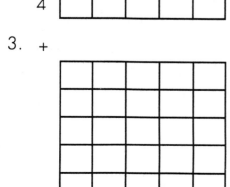

x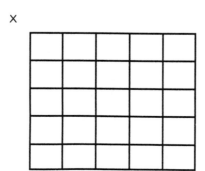

Quilt Design II

Math Topic: Tessellations

Student Activity: "Tessellating Quilt Task Card"—In small groups or partners, students tessellate a shape to create a quilt design. Groups need rulers, scissors, tape, and graph or dot paper.

Background

Quilt designs are often tessellations of a basic shape. A *tessellation* is a design that repeats itself over a flat surface without overlapping or leaving gaps. All triangles and quadrilaterals will tessellate. Shapes will tessellate if, when the shapes are placed next to each other around a point, the sum of the angles around the point is 360°.

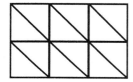

Triangles

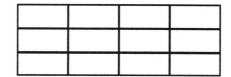

Rectangles

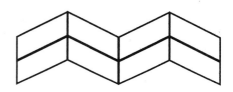

Other

One method to create a tessellation pattern:

1. Start with a rectangle.

2. Cut out a shape and place it elsewhere on the rectangle to make a figure.

3. Step 2 can be repeated to make a more elaborate design.

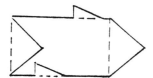

4. Repeat the design to complete the tessellation pattern.

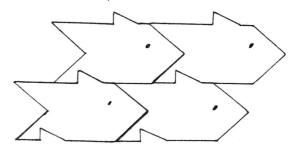

Tessellating Quilt
Task Card

Task: Design a quilt pattern by tessellating an irregular shape.

Materials: Ruler, scissors, tape, graph or dot paper

Background:

Quilt designs are often tessellations of some basic shape. A tessellation is a design that repeats itself over a flat surface without overlapping or leaving gaps. All triangles and quadrilaterals will tessellate.

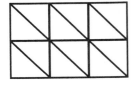

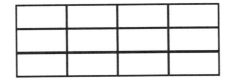

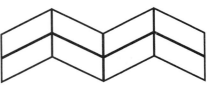

Triangles   Rectangles   Other

One method to create a tessellation pattern:

1. Start with a rectangle.

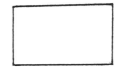

2. Cut out a shape and place it elsewhere along the rectangle to make a figure.

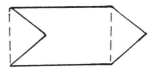

3. Step 2 can be repeated to make a more elaborate design.

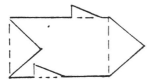

4. Repeat the design to complete the tessellation pattern.

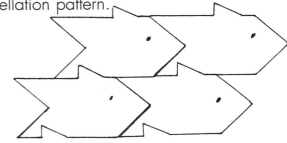

Follow steps 1 through 4 to create your own tessellation for a quilt design.

1. Draw a rectangle on graph or dot paper.

2. Cut out a shape and place it elsewhere along the rectangle to make a figure.

3. Repeat step 2 to make a more elaborate design.

4. Repeat the design to complete a tessellation approximately 20 cm x 20 cm.

Radio

Student Activity: "Up and Down the Dial"—Students calculate frequency and wavelengths of radio waves. Students use metric prefixes to make necessary conversions.

Background

Radio waves are part of the electromagnetic spectrum and therefore travel at the speed of light. Radio waves range from 100,000 m to 0.1 cm in length. The wavelength and frequency are characteristics that describe the alternating current signal. The wavelength of one cycle is the distance between an initial reference point such as zero amplitude and the return to the reference point. *Frequency*, measured in hertz (Hz), is the number of complete cycles generated in one second. Radio frequencies are measured in terms of thousands, millions, and billions of hertz.

Kilohertz (kHz)—1,000 Hz
Megahertz (MHz)—1,000,000 Hz or 1,000 kHz
Gigahertz (GHz)—1,000,000,000 Hz or 1,000MHz

Since radio waves travel at the speed of light 300,000,000 meters/sec, the frequency can be determined by dividing 300,000,000 by the wavelength in meters.

$$f = \frac{300{,}000{,}000}{w} \quad Hz$$

Note: Frequencies must be in hertz to use the given equation.

Fascinating Fact

Sound waves travel about 335 m/sec. compared to radio waves that travel 300,000,000 m/sec.

Band Designations:	Frequency Range:	Wavelength:
VLF—Very Low	3 to 30 kHz	100,000 to 10,000 m
LF—Low	30 to 300 kHz	10,000 to 1,000 m
M—Medium	300 to 3,000 kHz	1,000 to 100 m
H—High	3 to 30 MHz	100 to 10 m
VHF—Very High	30 to 300 MHz	10 to 1 m
UHF—Ultra High	300 to 3,000 MHz	1 to 0.1 m
SHF—Super High	3 to 30 GHz	10 to 1 cm
EHF—Extremely High	30 to 300 GHz	1 to 0.1 cm

Up and Down the Dial

Radio waves are part of the electromagnetic spectrum and range from 100,000 m to 0.1 cm in length. The *wavelength* is the distance between cycles of the wave. *Frequency*, measured in hertz (Hz), is the number of complete cycles generated in one second. Radio frequencies are measured in terms of thousands, millions, and billions of hertz.

Kilohertz (kHz)—1,000 Hz

Megahertz (MHz)—1,000,000 Hz

Gigahertz (GHz)—1,000,000,000 Hz

Since radio waves travel at the speed of light (300,000,000 meters/sec.), the frequency can be determined by dividing 300,000,000 by the wavelength in meters.

$$f = \frac{300,000,000}{w} \text{ Hz}$$

Note: Frequencies must be in hertz. Wavelength must be in meters.

Solve the following:

1. Find the frequency of a radio signal with a wavelength of 6 cm.
   a. Convert the wavelength to meters. _____
   b. Use the formula to calculate the frequency. _____
   c. Convert to Gigahertz. _____

2. Find the wavelength of a radio wave from a transmitter operating at 7.25 MHz.
   a. Write the formula for w. _____
   b. Convert the frequency to hertz. _____
   c. Substitute for f and find w. _____meters

3. Most novice radio communications are permitted in the VHF (Very High) and UHF (Ultra High) bands. The VHF band has wavelengths from 1 to 10 m. The UHF band has a frequency range of 300 to 3,000 MHz.
   a. Find the frequency range for the VHF band. _____ to _____
   b. Find the wavelength range for the UHF band. _____ to _____

Fascinating Fact | *Sound waves travel about 335 m/sec compared to radio waves that travel 300,000,000 m/sec.*

Sound

Student Activity: "Sound Off"—Students use formulas to determine frequency, speed, period, and wavelength of sound waves.

Background

Sound is caused by vibrations. In order to hear a sound, it must be transmitted through matter: solid, liquid, or gas. Sound cannot travel in a vacuum. Sound travels 1,100 ft./sec. in air, 4,700 ft./sec. in water, and 16,400 ft./sec. in steel.

Frequency refers to the number of sound waves that pass a location per second. The unit for frequency is the *hertz* named after Heinrich Hertz. The symbol is Hz and stands for vibrations per second. The period is the time in seconds it takes one vibration to pass a location. The product of the frequency and the period is 1; therefore, the frequency and period of a sound wave are reciprocals. As one increases, the other decreases.

The greater the frequency, the higher the pitch of the sound. The *pitch* refers to the highness or lowness of a sound. *Amplitude* refers to the loudness or softness of a sound. If you lightly pluck a violin string, you will hear the same note as when you pluck it more vigorously and it sounds louder. The pitch is the same, the amplitude has changed.

Wavelength is the distance between pulses. The speed of a sound wave is the product of its frequency and its wavelength.

$$\text{frequency (Hz)} = \frac{\text{vibrations}}{\text{second}}$$

$$\text{period (sec.)} = \frac{1}{\text{frequency (Hz)}}$$

speed (ft./sec.) = frequency (waves/sec.) x wavelength (ft./wave)

Sound Off

Sound is caused by vibrations. In order to hear a sound, it must be transmitted through matter. Sound cannot travel in a vacuum. Sound travels approximately 1,100 ft./sec. in air, 4,700 ft./sec. in water, and 16,400 ft./sec. in steel.

The unit for frequency is the *hertz* (Hz) named after Heinrich Hertz and stands for vibrations per second. The period is the time in seconds it takes one vibration to pass a location. *Wavelength* is the distance between pulses. The speed of a sound wave is the product of its frequency and its wavelength.

$$\text{frequency (Hz)} = \frac{\text{vibrations or waves}}{\text{second}} \qquad \text{period (sec.)} = \frac{1}{\text{frequency (Hz)}}$$

$$\text{speed (ft./sec.)} = \text{frequency (waves/sec.)} \times \text{wavelength (ft./wave)}$$

Use the given formulas to solve the following problems.

1. A piano string vibrates 50 times a second. Find the period of the note. _____

2. An oboe note has a period of 0.015625 sec. Find the frequency. _____

3. As the frequency increases, the period _____

4. What is the product of the frequency and the period of a note? _____

5. What is the speed of a 200 Hz sound wave with a 5.4 ft. wavelength? _____

6. What is the wavelength of a 440 Hz sound wave traveling 1,078 ft./sec.? _____

7. What is the frequency of a ¼ ft. sound wave traveling at 1,100 ft./sec.? _____

8. Hummingbirds vibrate their wings at 90 Hz. What is the period of the resulting sound wave? _____

Fascinating Fact | *The highest audible sound has a frequency of 20,000 Hz.*

Sun and Moon

Student Activity: "Sun and Moon"—Students solve problems using data concerning the sun and moon.

Background

The activity has students solve problems to determine the radii of the sun and the moon. Students use the results to calculate the average distance to the moon.

When the moon passes between the earth and the sun, a solar eclipse happens. On part of the earth, the eclipse may be total. The diagram (not to scale) shows the relationship during a solar eclipse.

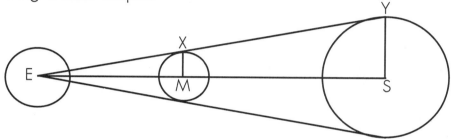

Since the moon totally eclipses the sun, common tangents may be drawn. The line passing through the centers of the earth, moon, and sun is the leg of the right triangles shown. The triangle EMX is similar to triangle ESY. The ratio of the parts of the small triangle equals the ratio of the parts of the large triangle. Therefore, MX/EM = SY/ES. The radius of the moon is to the distance of the moon from the earth as the radius of the sun is to the distance of the sun from the earth.

Fascinating Fact

If a soccer ball is used to represent the sun, the earth would be a BB a half-mile away.

DATA:
(Distances
are approx.)

Diameter of the earth	8,000 mi.
Radius of the earth	4,000 mi.
Diameter of the sun	880,000 mi.
Radius of the sun	440,000 mi.
Diameter of the moon	2,160 mi.
Radius of the moon	1,080 mi.
Distance from earth to sun	93,000,000 mi.
Distance from earth to moon	230,000 mi.
Surface area of earth	64,000,000 π sq. mi. or
	201,000,000 sq. mi.

Name _____
Sun and Moon

Solve the following problems concerning the sun and moon.

1. The sun is 110 times wider than the earth which has a diameter of about 8,000 miles. Find the diameter of the sun. _____

   Find the radius of the sun. _____

2. If the moon were placed on top of the United States, it would cover 90% of the distance from New York to Los Angeles (2,400 miles). Find the diameter of the moon. _____

   Find the radius of the moon. _____

3. When the moon comes directly between the earth and the sun, a solar eclipse occurs. Use the figure (not to scale) to write a proportion involving the distances from the earth to the sun and to the moon and the radii of the sun and moon. _____

   _____

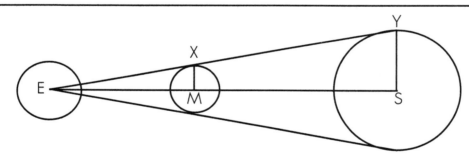

4. The sun is 93,000,000 miles (approx.) from the earth. Use the proportion from #3 and the radii from #1 and #2 to find the distance from the earth to the moon (round to nearest thousand miles). _____

5. Surface area of a sphere equals the area of four great circles: S.A. = $4\pi r^2$. Find the surface area of the earth to the nearest million sq. mi. _____
   Two thirds of the earth is covered by water. How many square miles are covered by water? _____

Fascinating Fact | *If a soccer ball is used to represent the sun, the earth would be a BB a half-mile away.*

Trajectories

Math Topics: Solving equations, squares, square roots

Student Activity: "What Goes Up Must Come Down"—Students determine the heights and times describing the trajectory of various projectiles.

Background

A *trajectory* is the path a thrown object called a *projectile* takes. The trajectory is in the shape of a parabola. In the seventeenth century, Sir Isaac Newton explained the motion of objects. Newton's Laws of Physics explain why the path is a parabola as opposed to a straight line or semicircle.

Newton's First Law states that an object at constant speed will travel in a straight line unless acted upon by an outside force. In the case of a thrown object the outside force is gravity. Newton's Second Law of Physics states that an object will accelerate in the direction of the force affecting it. The force of gravity is toward the center of the earth, downward. No force acts on the object in the horizontal direction, so it will travel at a constant speed horizontally. (Air resistance may play a small part in slowing the horizontal motion of nonaerodynamic objects.) Gravity acts on the vertical motion of the object causing it to decelerate in the upward direction until it stops and starts to accelerate downward.

The trajectory of an object depends on the initial velocity of the projectile and the angle at which it was thrown or launched. *Ballistics* is the study of projectiles and their related trajectories. The study of ballistics started in the 1500s by Niccolo Tartaglia, but the equipment to measure the initial velocity of an object was not developed until the 1700s by Benjamin Robins.

A basic trajectory formula for the height of the projectile is

$$h = -16t^2 + v_o t + h_o$$

where h is the height in feet, t is the time in seconds, v_o is the initial velocity, and h_o is the initial height.

What Goes Up
Must Come Down

T

A *trajectory* is the path a thrown object called a *projectile* takes. The trajectory is in the shape of a parabola. The force of gravity causes the object to decelerate during its upward motion and then causes it to accelerate downward.

Solve the following problems.

1. Graph the height of a projectile (fired from ground level) versus time that is defined by
$$h = -16t^2 + 96t$$
where h is in feet and t is in seconds.

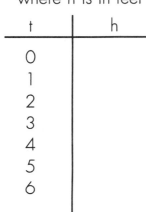

t	h
0	
1	
2	
3	
4	
5	
6	

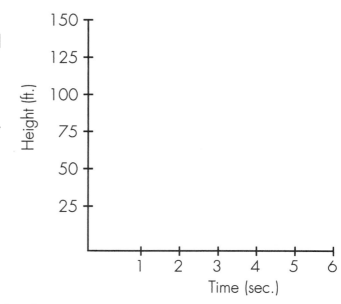

a. Give the height after 3 seconds. _____
b. What happens at 3 seconds? _____
c. Will the projectile reach 200 feet? _____
d. What is true after 2 seconds and 4 seconds? _____

2. A rocket is fired into the air. Its height is defined by the equation:
$$h = -16t^2 + 64t + 2,240$$
a. Its maximum height is reached after 2 sec. Find the height. _____
b. Find the height after 1 second. _____
c. Find the height after 3 seconds. _____
d. Explain the results from b & c. _____

3. A rock is thrown upward from the top of a 60-story building (880 feet) with an initial velocity of 96 ft./sec. Its height is described by
$$h = -16t^2 + 96t + 880$$
a. Find the heights after 2 seconds, 3 seconds, and 4 seconds. _____
_____
b. What occurs at 3 seconds? _____
c. Find the height after 11 seconds. _____
d. What does the answer for c mean? _____

Uranium

Student Activity: "Uranium Decay"—Students solve problems pertaining to exponential growth and decay.

Background

Uranium-238 is a radioactive element that decays to Lead 206, a stable element. A number of isotopes are produced in the decay process. The time it takes for one half of the unstable element (the parent element) to decay to atoms of a more stable form (the daughter element) is called the *half-life* of the radioactive element. The half-life of a given element is a constant. The half-lives of various elements range from a billionth of a second to 49 billion years. Uranium-238 (daughter lead-206) has a half-life of 4.5 billion years. Uranium-235 (daughter lead-207) has a half-life of 704 million years.

Radioactive decay occurs at a geometric rate. Therefore, the graph of the decay rate is a curve, not a straight line. The rate of decay is ½. An element with 1,000,000 parent atoms will have 500,000 (½ of 1,000,000) parent atoms and 500,000 daughter atoms after one half-life. After two half-lives, the number of parent atoms will be 250,000 (½ of ½ of 1,000,000) and the number of daughter atoms will be 750,000. By measuring the parent to daughter ratio and knowing the half-life of the parent, a geologist can determine the age of the sample.

Exponential growth occurs when the rate is greater than one. The number of cells of a certain virus might double every day. One thousand cells become 2,000 cells after one day and 4,000 cells after two days. The graph is, therefore, a curve and not a straight line.

Uranium Decay

Uranium-238 is a radioactive element that decays to Lead-206, a stable element. The time it takes for one half of the unstable element to decay to atoms of a more stable form is called the *half-life* of the radioactive element. Uranium-238 has a half-life of 4.5 billion years. Uranium-235 has a half-life of 704 million years.

1. Assume there are 100 unstable atoms of a radioactive element. Graph the number of unstable atoms after each half-life.

Half-life Unit	Unstable Atoms
0	100
1	50
2	
3	
4	
5	

2. A rock contains one fourth of the number of Uranium-235 elements that a "new" sample would have. How old is the rock? _____

3. Assume that a bacteria doubles every day. Graph the number of bacteria cells each day.

Days	Bacteria Cells
1	100
2	200
3	400
4	
5	
6	

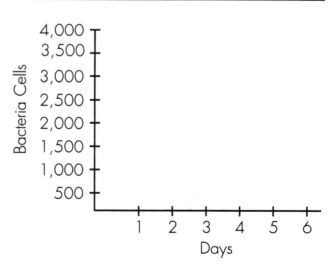

4. A bacteria sample of cells that double every day contains 10,000 cells. How many cells were there yesterday? _____
   Will there be tomorrow? _____

Violin

Math Topics: Matrix, ratios, graphing, slope

Student Activity: "Virtuoso Violins"—Students compare measurements of parts of violins, violas, cellos and basses by computing ratios. Students then graph data from a matrix and compare the slopes of the resulting lines to the calculated ratios.

Background

The violin branch of the strings family of orchestra instruments contains the violin, viola, cello, and bass. The size of the instrument determines the tone of the instrument—the smaller the instrument, the higher the tone and the larger the instrument, the deeper or lower the tone. A viola has a deeper tone than a violin. The tone of a cello is deeper than a viola, and a bass has the deepest, lowest, tone.

The instruments are available in smaller sizes for children to use. Violins are available in 4/4 (full size), 7/8, 3/4, 1/2, 1/4, 1/8, 1/16, 1/32, and 1/64 sizes. Violas come in large and small sizes. Cellos and basses come in 4/4, 3/4, 1/2, 1/4, and 1/8 sizes. The fractions do not represent fractional parts. A 1/2 violin is not half the size of a full violin.

The measurements in the matrix are given in inches.

Matrix of Sizes:

	Violin	Viola	Cello	Bass
neck	9.7	11.0	18.9	30.9
fingerboard	12.8	14.5	27.0	43.3
body	14.1	16.0	29.9	45.7
length	23.2	26.4	48.8	74.8

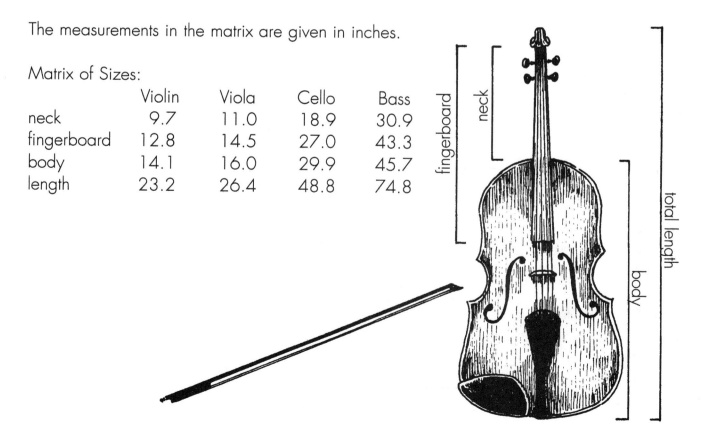

Virtuoso Violins

The violin branch of the strings family of instruments contains the violin, viola, cello, and bass. The size of the instrument determines the tone of the instrument. The violin has the highest tone. The bass has the deepest, lowest, tone.
Matrix of Sizes:

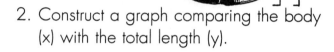

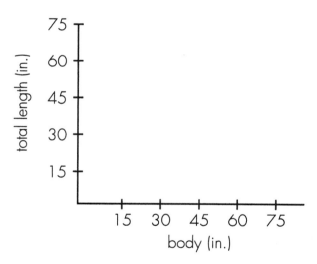

	Violin	Viola	Cello	Bass
neck	9.7	11.0	18.9	30.9
fingerboard	12.8	14.5	27.0	43.3
body	14.1	16.0	29.9	45.7
length	23.2	26.4	48.8	74.8

1. Construct a graph comparing the neck (x) with the fingerboard (y) measurements.

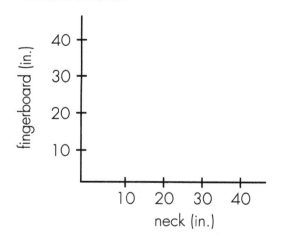

2. Construct a graph comparing the body (x) with the total length (y).

3. Compute the slope of the line through the points in graph 1.

4. Compute the slope of the line through the points in graph 2.

5. Compute the ratio of the fingerboard to the neck for each instrument.

Violin _____
Viola _____
Cello _____
Bass _____

6. Compute the ratio of the length to the body for each instrument.

Violin _____
Viola _____
Cello _____
Bass _____

Wave Motion

Math Topics: Rewriting formulas, solving formulas

Student Activity: "Ride the Wave"—Students solve problems using information and formulas pertaining to waves and wave motion.

Background

The highest point of a wave is the *crest*, and the lowest point of a wave is the *trough*. The distance between two crests or between two troughs is called the *wavelength*. The vertical distance between the crest and the trough is the *wave height* or *wave depth*.

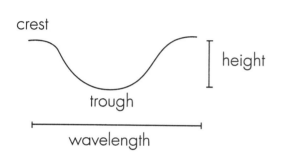

$$\text{wave depth} = \frac{\text{wavelength}}{2}$$

Waves will usually break when the water depth is about 1.3 times the wave height.

$$\text{water depth} = 1.3 \times \text{wave height}$$

The *period* of a wave is the time it takes a wave from crest to crest to pass a point.

$$\text{period (s)} = \frac{\text{wavelength (ft.)}}{\text{speed of wave (ft./s)}}$$

Ocean waves are *transverse waves*. The movement of the wave is perpendicular to the motion the wave creates. An object floating in water actually moves in a slight circular motion as the wave pattern moves forward. The diameter of the circular path is the height of the wave. X-rays, radio waves, and microwaves are also transverse waves.

Fascinating Fact

Swells, ocean waves, can stay about the same distance apart for thousands of miles.

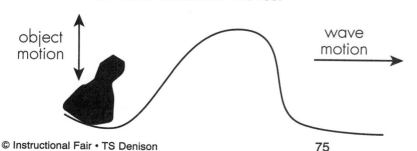

Ride the Wave

Use the following diagram and formulas to solve the following problems. Complete the crossnumber puzzle.

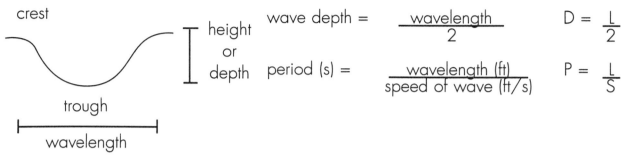

crest

height
or
depth

trough

wavelength

wave depth = $\dfrac{\text{wavelength}}{2}$ $D = \dfrac{L}{2}$

period (s) = $\dfrac{\text{wavelength (ft)}}{\text{speed of wave (ft/s)}}$ $P = \dfrac{L}{S}$

ACROSS

1. The distance between the crest and the trough of a wave is 17.5 ft. Find the wavelength.

2. Find the wavelength of an ocean wave that has a period of 15 s and a speed of 76 ft./s.

4. Find the wave depth if the wavelength is 1.16 m.

6. Waves will break when the water depth is approximately 1.3 times the height of the wave. At what depth will a wave 38 ft. high break?

8. Waves are breaking in water 3.25 m deep. Find the wave height.

DOWN

1. Find the wave depth of a wave that has wavelength of 60 ft.

2. Find the wavelength of a wave with a wave depth of 52 ft.

3. Find the speed of a wave with a period of 8 s and a wavelength of 340 ft.

5. The horizontal distance between a crest and a trough is 40.2 ft. Find the wavelength.

7. At what depth will a wave 7.5 m high break?

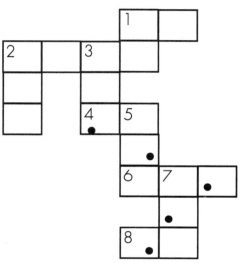

Wind

Student Activity: "Blowing in the Wind"—Students substitute Beaufort scale numbers into a given equation to determine wind speed. The correct answers result in spelling *Francis Beaufort*, the developer of the scale.

Background

The Beaufort scale is a scale from 0–12 used to indicate wind speed. It was developed by Admiral Sir Francis Beaufort in 1805. It uses easily observable phenomena such as smoke rising, tree branch movement, and flag waving.

Beaufort Scale:

0	calm air	smoke rises vertically
1	light air	smoke shows wind direction
2	light breeze	wind can be felt
3	gentle breeze	flags extend, leaves are blown
4	moderate breeze	small tree branches move
5	fresh breeze	flags ripple, small trees sway
6	strong breeze	large tree branches move
7	near gale	large trees sway
8	gale	twigs break
9	strong gale	roof damage
10	storm	trees uprooted
11	violent storm	widespread damage
12	hurricane	devastation

The relationship between the Beaufort scale number (s) and the wind speed (w) in miles per hour can be approximated by the equation: $s = 2\sqrt{(w + 9)} - 6$. Given the observable effect of the wind, one can substitute the Beaufort number into the equation and solve for w. Steps:

1. Substitute for s and add 6.
2. Divide the result by 2.
3. Square the result.
4. Subtract 9 from the square.
5. Round result to nearest mile per hour.

Fascinating Fact

A knot is a unit of speed equal to 1.15 miles per hour (one nautical mile per hour). To convert miles per hour to knots, divide by 1.15.

Blowing in the Wind

In 1805 a scale based on the observable effects of wind was devised. The scale can be used to determine the wind speed.

$S = 2\sqrt{w + 9} - 6$ relates the scale number (S) to the approximate wind speed (w) in miles per hour.

Substitute the scale numbers into the equation, and solve for the wind speed. Round to the nearest mile per hour. Place the letter of the wind speed above the scale number to discover the name of the developer of the scale.

Answer Key:

(w in miles per hour)

			Scale (s):	
A = 11	B = 40	_____	0	smoke rises vertically
C = 21	D = 23	_____	1	smoke shows wind direction
E = 47	F = 3	_____	2	wind can be felt
G = 48	H = 18	_____	3	flags extend, leaves are blown
I = 27	J = 29	_____	4	small tree branches move
K = 49	L = 15	_____	5	flags ripple, small trees sway
M = 60	N = 16	_____	6	large tree branches move
O = 63	P = 70	_____	7	large trees sway
Q = 15	R = 7	_____	8	twigs break
S = 33	T = 72	_____	9	roof damage
U = 55	V = 75	_____	10	trees uprooted
W = 50	X = 9	_____	11	widespread damage
Y = 59	Z = 10	_____	12	devastation

$$\overline{1}\ \ \overline{2}\ \ \overline{3}\ \ \overline{4}\ \ \overline{5}\ \ \overline{6}\ \ \overline{7}\qquad \overline{8}\ \ \overline{9}\ \ \overline{3}\ \ \overline{10}\ \ \overline{1}\ \ \overline{11}\ \ \overline{2}\ \ \overline{12}$$

Fascinating Fact | *A knot is a unit of speed equal to 1.15 miles per hour (one nautical mile per hour). To convert miles per hour to knots, divide by 1.15.*

Xerography

Math Topics: Ratios, percent, similarity, area formulas

Student Activity: "Copycat"—Students answer questions concerning enlargements and reductions of documents. Linear measures and areas of the enlargements and reductions are calculated. The answers result in spelling *xerography*.

Background

The fundamental principles of electrophotography, later named xerography, were defined in a patent application in 1939. The process uses the fact that when light strikes a photoconductive material, the electrical conductivity of the material increases.

In most common photocopiers, the original document is moved from a document handler to the platen where it is projected by lamps, mirrors, and lenses onto the photoreceptor belt. The charge of static electricity from the belt is discharged to the areas receiving light from the projected image. This forms a latent image. Magnetic rollers brush the belt with dry ink of opposite charge. The dry ink clings to the latent image on the photoreceptor. The image becomes visible. The copy paper moves to the belt. The static electricity charge of the paper attracts the dry ink from the belt. Heat and pressure fuses the dry ink into the paper. The copy paper is ejected from the machine.

The ratio of the linear measurements of the copy to the original is the percent of enlargement or reduction. Example, the 80% setting on the machine means a 10-inch original will result in an 8-inch copy. Likewise, a 120% setting results in a 12-inch copy. The area of an 80% copy will be 0.64 (0.8 x 0.8) of the area of the original document. The area of a 120% copy will be 1.44 (1.2 x 1.2) of the area of the original document.

Copycat

Answer the following questions to determine the name given to electrophotography.
Round each answer to the nearest hundredth or percent.

1. What is the length of a 75% copy when the original document is 11 inches long?
   A. 6.82 in.   E. 8.25 in.   I. 14.67 in.

2. An 8½" x 11" document has to be reduced to a 6" x 8" card. What percent setting should be used so that the entire document is on the smaller card?
   E. 80%   I. 75%   O. 71%

3. What is the width of a 120% copy if the original is 9 inches wide?
   R. 10.8 in.   S. 7.5 in.   T. 13.33 in.

4. A 5-inch line is reduced to a 4-inch line. What was the percent setting?
   N. 125%   O. 20%   P. 80%

5. What is the width of a 65% copy of an 8½" x 11" inch original?
   X. 7.65 in.   Y. 5.53 in.   Z. 7.15 in.

6. An 8-inch original is enlarged to 10 inches. What was the percent setting?
   F. 80%   G. 25%   H. 125%

7. What is the area of a 9" x 12" document?
   A. 108 sq. in.   B. 42 sq. in.   C. 54 sq. in.

8. What is the area of a 75% copy of an 8" x 10" original?
   E. 80 sq. in.   F. 36 sq. in.   G. 45 sq. in.

9. An 8" x 10" original becomes a 10" x 12.5" copy. What was the percent setting?
   P. 80%   Q. 156%   R. 125%

10. What setting should be used to enlarge 8 inches to 12 inches?
    X. 150%   Y. 75%   Z. 125%

$$\overline{10} \quad \overline{1} \quad \overline{9} \quad \overline{2} \quad \overline{8} \quad \overline{3} \quad \overline{7} \quad \overline{4} \quad \overline{6} \quad \overline{5}$$

X-rays

Student Activity: "X-ray Vision"—Students answer questions pertaining to electromagnetic waves including X-rays. The answers will spell out *Wilhelm Roentgen*, the discoverer of X-rays.

Background

X-rays are a type of electromagnetic radiation with a frequency higher than visible light but lower than gamma rays. Initially X-rays were measured in X units (XU) based on the structure of rock salt. More recently, X-ray wavelengths have been compared to the meter. The original XU was to be equivalent to 10^{-11} cm. The actual quantity was 0.202% larger. It is customary to give wavelengths in terms of Angstroms Å which equals 10^{-8} cm. Therefore, 1,000 XU = 1.00 Å, to the nearest 0.01 Angstrom.

In 1895 Wilhelm Roentgen discovered some properties of X-rays, such as that soft body tissue was transparent to the rays, but that bones blocked them. He did not know what the rays were so he called them *X-rays*. They are produced when electrons are accelerated to high speeds by means of potential differences of 20,000 volts. When the electrons hit matter, the kinetic energies are converted into very high frequency electromagnetic waves (X-rays).

X-rays are produced in cathode-ray tubes, such as TV picture tubes. The faceplate glass contains lead to stop the X-rays and protect the viewer.

X-ray Vision

According to Dell Comics, Superman has X-ray vision, which enables him to see through doors and walls. In reality, X-rays will pass through soft body tissue, are stopped by bone, and cannot penetrate lead.

X-rays are produced when electrons are accelerated to high speeds. When the electrons hit matter, the kinetic energies are converted into very high frequency electromagnetic waves (X-rays).

Solve the following problems. Cross out the correct answers below to determine who discovered X-rays.

1. X-rays were originally measured in X units (XU). 1 XU = 10^{-11} cm. How many XUs are in one Angstrom if 1Å equals 10^{-8} cm?

   _____

2. It takes a potential difference of at least 2×10^4 volts to generate X-rays. Write the number in standard form.

   _____

3. The frequency of X-rays are in the 10^{19} Hz range. Visible light is in the 10^{15} Hz range. How many times faster do X-rays vibrate?

   _____

4. X-rays were discovered 230 years after Newton's experiments with light in 1665. When were X-rays discovered?

   _____

5. Light waves are 10^{-6} m. Convert the wavelength to centimeters.

   _____

6. X-rays are 10^{-10} m. Convert the wavelengths to centimeters.

   _____

7. Gamma radiation, emitted from nuclear reactions, has wavelengths of less than 10^{-13} m. Convert to cm.

   _____

8. 10^3 volts equals 1 kV. Convert 2×10^4 volts to kV.

   _____

RO	WI	L	LH	LI	EL	A	M
10^{-11}	0.001	1,000	10^{-15}	10^4	2	10^{-4}	10^2
ST	RO	TH	EN	ER	TG	S	EN
20,000	10^{-12}	1895	10^{-2}	10^{-8}	200	20	1785

___ ___ ___ ___ ___

___ ___ ___ ___ ___

Year

Student Activity: "Days: Past, Present, and Future"—Students follow directions to calculate the days of various dates. The problems provide historical information. The correct answers spell out "algorithm."

Background

An *algorithm* is a rule or process for solving a problem. The following is an algorithm, and example, for finding the day of the week for any date from 1900.

Example: June 11, 1951
1. Write down the difference of the year and 1900.   1951 – 1900   1. 51
2. Divide the difference by 4 (drop any remainder). $^5\!\!\frac{1}{4}$ = 12 r 3   2. 12
3. Write the month number from the Month Chart.   June—5   3.   5
4. Write the date of the month.   11   4. 11
5. Add the results from 1 – 4.   51+12+5+11 = 79   5. 79
6. Divide the sum by 7. Write the remainder. $^{79}\!\!/\!_7$ = 11 r 2   6.   2
7. Find the day in the Day Chart.   Monday

The division by 4, step 2, takes into account the number of years the date is beyond a leap year. The Month Chart numbers depend on the number of days in a given year and the number of days in the given month. The Day Chart numbers are the possible remainders when dividing by 7, the number of days in a week.

Fascinating Fact

Triskaidekaphobia is the fear of 13. Friday the thirteenth happens 48 times every 28 years.

Month Chart:
January—1 (0 leap yr.)	February—4 (3 leap year)
March—4	April—0
May—2	June—5
July—0	August—3
September—6	October—1
November—4	December—6

Day Chart:
Sunday—1	Monday—2
Tuesday—3	Wednesday—4
Thursday—5	Friday—6
Saturday—0	

Days: Past, Present, and Future

The current calendar has 12 months, with 365 days a year (366 in leap years) divided in 52+ weeks. Use the following steps to calculate the day of the week for the dates given in the following problems.

1. Write down the difference of the year and 1900.
2. Divide the difference by 4 (drop any remainder).
3. Write the month number from the Month Chart.
4. Write the date of the month.
5. Add the results from 1 – 4.
6. Divide the sum by 7. (Do not use a calculator.) Write the remainder.
7. Find the day in the Day Chart.

MONTH CHART
January—1 (0 leap yr.)
February—4 (3 leap yr.)
March—4
April—0
May—2
June—5
July—0
August—3
September—6
October—1
November—4
December—6

DAY CHART
Sunday—1
Monday—2
Tuesday—3
Wednesday—4
Thursday—5
Friday—6
Saturday—0

1. John F. Kennedy was born on May 29, 1917.
   GO) Tuesday   MA) Thursday

2. John Glenn was the first American to orbit earth on February 20, 1962.
   SH) Monday   TH) Tuesday

3. June 16, 2063, will be the 100th anniversary of Valentina Tereshkova being the first woman in space.
   LA) Sunday   AL) Saturday

4. July 4, 2001, will be the two hundred twenty-fifth anniversary of Independence Day.
   M) Wednesday   N) Thursday

5. Valentine's Day February 14, 2007
   RI) Wednesday   RA) Friday

Bonus: Your twenty-first birthday: _____
         Your fiftieth birthday: _____

A rule or process for solving a problem is called: __ __ __ __ __
                                                    3   1   5   2   4

Fascinating Fact

Triskaidekaphobia is the fear of 13. Friday the thirteenth happens 48 times every 28 years.

Zero—Absolute Zero

Math Topics: Graphing data points, solving equations

Student Activity: "Absolutely Zero Task Card"—In small groups or partners, students graph data points and draw lines to estimate the absolute zero (0° K) equivalent on the Celsius scale. Students then solve equations to determine the absolute zero value (temperature at which all molecular motion stops) on the Celsius scale.

Background

The two most common temperature scales are Fahrenheit and Celsius (Centigrade). Scientists use another temperature scale called the Kelvin scale. Absolute Zero (0° K) on the *Kelvin scale* is the total absence of heat. Since molecular activity ceases to exist at absolute zero, the temperature cannot be measured directly. The equivalent temperature on the Celsius scale can be determined by measuring the volumes of various gases at different Celsius temperatures.

methane	(330.8°C, 4 L)
	(179.8°C, 3 L)
nitrous oxide	(350.8°C, 1 L)
	(38.8°C, 0.5 L)
water	(182.8°C, 2 L)
	(68.8°C, 1.5 L)

The linear relationships can be extended into volumes and temperatures at which the compounds are no longer gases. By substituting 0 for the volume (no molecular action), the Celsius equivalent can be determined.

> ### Fascinating Fact
>
> *Scientists use the Kelvin scale when studying the pressure, volume, and temperature relationships of gases defined by Boyle's and Charles' Gas Laws. The combined gas laws can be expressed by*
>
> $$\frac{P_1 V_1}{T_1} = \frac{P_2 V_2}{T_2}$$

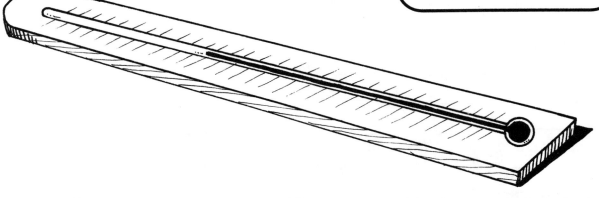

Absolutely Zero
Task Card

Z

Task: Determine the Celsius temperature for *absolute zero.*

Background:

Absolute Zero (0° K) on the Kelvin temperature scale is the total absence of heat. Since molecular activity stops at Absolute Zero, the temperature cannot be measured directly. The equivalent temperature on the Celsius scale can be determined by measuring the volumes of various gases at different Celsius temperatures. The linear relationships can be used to determine the temperature at 0 volume, no molecular action.

Activity:

1. Draw a coordinate graph.

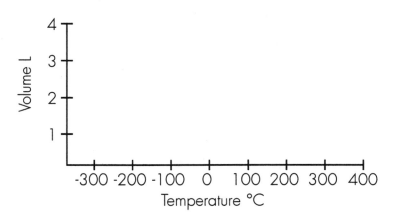

DATA	
compound	(temperature, volume)
methane	(330.8°C, 4 L)
	(179.8°C, 3 L)
nitrous oxide	(350.8°C, 1 L)
	(38.8°C, 0.5 L)
water	(182.8°C, 2 L)
	(68.8°C, 1.5 L)

2. Plot the data (temperature, volume) for each compound.

3. Draw a line through the points for each compound.

4. Estimate the point of intersection of the three lines. The point of intersection (when the volume is 0) of the three lines is absolute zero. _____

5. Choose one of the following three equations:

methane	$-2T + 302V = 546.4$
nitrous oxide	$-3T + 1872V = 819.6$
water	$-T + 228V = 273.2$

6. Let V = 0, solve the equation for T. So _____°C = 0°K

Fascinating Fact	*Scientists use the Kelvin scale when studying the pressure, volume, and temperature relationships of gases defined by Boyle's and Charles' Gas Laws.*

Zoo Animals

Math Topics: Ratios, formulas, data analysis, solving equations

Student Activity: "It's a Zoo Out There"—Students solve problems using data, facts, and formulas pertaining to various zoo animals.

Background

Facts about animals can provide some interesting math relationships. Students could research additional animal facts and write their own questions.

The ratio of food a mammal needs to its body mass varies by species. For a mouse, the food to body mass is 0.6. A mouse eats three fifths of its weight every day. For a moose it is 0.06. A moose requires three fiftieths of its weight. An elephant has a food to weight ratio of 0.02 or one fiftieth. Discussing the ratios for a mouse and a moose or an elephant and the resulting quantity of food emphasizes that a ratio is the *comparison* of two quantities and not the quantities themselves.

The ratio of the tail of an iguana to the rest of its body is 1.8 to 1.

A cheetah is the fastest land mammal and can reach a speed more than 60 mph (88 ft. per second) but can maintain its top speed for less than a minute. A gazelle can run 73 feet per second for several minutes. Gazelles can survive in the wild because they can sense the safe distance needed to outrun a cheetah.

> **Fascinating Fact**
>
> Koala *is Aborigine for "it does not drink."*

The number (N) of times per second a bird flaps its wings in normal flight is inversely proportional to its wing (W) length in cm. N x W = 216. The smaller the bird, the faster a bird must flap its wings. The larger the bird, the slower a bird must flap its wings.

Name _____

It's a Zoo Out There

Solve each problem. Place the letter associated with the correct answer above the problem number.

The ratio of food a mammal needs to its body mass varies by species. For a mouse it is 0.6. For a moose it is 0.06. A mouse that weighs 400 g must eat ____(1)____ g, whereas a moose that weighs 400 kg needs ____(2)____ kg.

The ratio of the tail of an iguana to the rest of its body is 1.8 to 1. If an iguana's tail measures 27 inches, the rest of its body would be ____(3)____ in. long. If an iguana's body (without the tail) is 30 cm, its tail is ____(4)____ cm long.

A cheetah is the fastest land mammal. It can reach a speed of more than 60 mph. Calculate the speed in feet per second ____(5)____ to determine how far a cheetah can run in 20 sec. ____(6)____ ft.

A gazelle can run 73 feet per second for several minutes. A gazelle must stay at least ____(7)____ feet from a cheetah to outrun it for 20 seconds.

The number (N) of times per second a bird flaps its wings in normal flight is inversely proportional to its wing (W) length in cm. N x W = 216. A gull with a 54 cm wing would flap its wings ____(8)____, whereas a bee hummingbird with a 3.6 cm wing would flap its wings ____(9)____.

ANSWER KEY:

A = 2,400	B = 6	C = 600
D = 1,760	E = 240	F = 80
G = 30	H = 150	I = 24
J = 1.5	K = 60	L = 3
M = 540	N = 88	O = 4
P = 3	Q = 3,000	R = 300
S = 15	T = 54	U = 8.8

Fascinating Fact "Koala" is Aborigine for

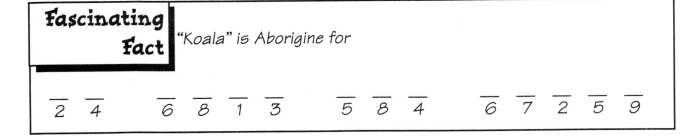

__ __   __ __ __ __   __ __ __   __ __ __ __ __
2 4   6 8 1 3   5 8 4   6 7 2 5 9

Answer Key

Amazing Atoms page 2
1. 9.1×10^{-25}
2. 1.67×10^{-21}
3. 2.2×10^{-8}
4. 4.8×10^{-11}
5. 1×10^{6}
6. 1×10^{15}
7. 3×10^{6}
8. 6.022×10^{23}

Atomic Numbers page 4

ELEMENT	ATOMIC NUMBER = P	ATOMIC MASS = P+N	P	E	N
1. Barium	56	137	56	56	81
2. Calcium	20	40	20	20	20
3. Krypton	36	83	36	36	47
4. Mercury	80	200	80	80	120
5. Silver	47	107	47	47	60
6. Sodium	11	23	11	11	12
7. Uranium	92	238	92	92	146
8. Zinc	30	65	30	30	35

Pythagorean Theorem of Baseball page 6
1. 114
2. 79
3. 100
4. 98
5. 94
Allowed more runs than runs made

Bouncing Balls Task Card page 8
Answers will vary. Data points will be on a line with a slope equal to the bounce difference divided by the drop difference.

Bowling page 10

1.

1	2	3	4	5	6	7	8	9	10
6 2	X	7 /	5 3	9 /	X	3 2	7 2	8 /	2 5
8	28	43	51	71	86	91	100	112	119

2.

1	2	3	4	5	6	7	8	9	10
8 /	X	7 /	X	9 /	7 1	X	X	X	6 / 9
20	40	60	80	97	105	135	161	181	200

3. All strikes

Calorie Count page 12
1. A
2. R
3. M
4. T
5. L
6. O
7. I
8. E
9. C
10. B
Bomb Calorimeter

Cricket Temperature page 14
1. 60°–62°F
2. 75°–77°F
3. 50°–52°F
4. 90–92 chirps/min.
5. 130–132 chirps/min.
6. 50–52 chirps/min.
Extension: $T = C/4 + 37$

From Rustling Leaves to Jet Planes page 16
1. 100
2. 110
3. 65
4. 1/100; 1,000
5. 10
6. 25 db

$$(110 - 100) \times 1000 \times 1/100 + (65 - 25) + 10$$
$$10 \quad \times \quad 10 + 40 \ + \ 10$$
$$150$$

Diamonds Are Forever page 18
1. corundum
2. between 5 and 9
3. 3
4. 6.5
5. gypsum or talc
6. BDAC
7. diamond will scratch glass

Equation Exercise Task Card page 20
Answers will vary.

B-52 Fallout Task Card page 22
7.8 sec.
$0 = -16t^2 + 975$
$7.8 = t$
Since
12.5
$\times 78$
975

7 sec.

Giants: Fact or Fiction page 24
1. 2.25
2. 10,125
3. 144
4. 288 sq. yds.
5. 207,360 lbs.
6. 3,456,000 cal.
7. .014 sq. yd. or 2 sq. in.
8. 1.7 cal.
9. no; Answers will vary.

Weighty Matters page 26
1. 127.4 lbs.
2. 30,360 lbs. or 15.18 tons, 1,080 lbs. or .54 tons
3. 50 lbs.
4. no, It would weigh 200 lbs.
5. Saturn: 53.5 lbs. Neptune: 59 lbs.
6. 150 lbs.

Heart Disease—Risky Business
page 28

1. .94 S
2. .85 R
3. .75 N
4. 1.03 D
5. .83 I
6. .92 E
7. .78 A
8. .97 Y
9. 1.06 O
10. .71 T
11. .95 C

Circled: 2. Miss Wilson, 4. Mr. Talley, 5. Mrs. Amera, 8. Miss Zostel, 9. Mr. Freeman
CORONARY ARTERY DISEASE

Interesting Investments
page 30

1. $570
2. $572.45
3. $219.35
4. Continuously $62.50 or $62.44
   $23.08 or $23.02
5. 12 yrs.
6. About $16,000

Japanese Magic Circle Task Card
page 32

1,326 Possible Method: Pairs total 53
Each diameter (-1) = 5 x 53 = 265
Times # of diameters x 5

```
    1,325
+       1   center
    1,326
```

Space Shuttle Task Card
page 34

1. 6.65×10^{-12}
2. 6550 km
3. 1.36 hr.

Konigsberg's Bridges Task Card
page 36

No
Networks with more than 2 odd vertices cannot be traced.

Network	#even	#odd	Traceable?
1	3	2	Yes
2	1	4	No
3	2	2	Yes
4	2	4	No
5	0	4	No
6	2	2	Yes
7	3	2	Yes
8	1	4	No

Reaching New Heights
page 38

1. 3 ft.
2. yes. It will reach 11.6 ft.
   $12^2 = 3^2 + H^2$
   $135 = H^2$
   $11.6 = H$
3. 14 ft.
   $15^2 = 5^2 + H^2$
   $200 = H^2$
   $14 = H$
4. 21' ladder
   $L^2 = (20/4)^2 + 20^2$
   $L^2 = 25 + 400$
   $L^2 = 425$
   $L = 20.6$
5. 75°

Storm Tracker Task Card
page 40

1. 982,080,000 ft./sec.
2. 982,000,000 ft./sec.
3. 9.82×10^8 ft./sec.
4. 4.8 sec.
5. 5 sec.
6. 5
7. 0.33 km/sec.
8. 3 sec.
9. 3

10. After seeing the lightning and before hearing the thunder, count five seconds for every mile or count three seconds for every kilometer.

Magic Card Task Card
page 42

Answers will vary.

Med-Math Shortcut
page 44

1. 54 kg
2. 63 kg
3. 32.4 kg
4. 72 kg
5. 900 kg
6. 4.5 kg
7. 225 kg
8. 18 kg
9. 36 kg
10. 153 kg

Extension: 99, 1%

Puttering Around
page 46

Diagrams may vary.

Chambered Nautilus: A Golden Shell
page 48

Answers will depend on actual measurements used. l = 1.6 w

Nutrition Labels: You Are What You Read
page 50

1. 29%
2. 52%
3. Soup 25 mg/serv
4. Stew
5. 60 mg
6. Stew 8g/serv
7. Vitamin C
8. Soup ⅜
   Stew 1 ⅕

Outdoor Orbits Task Card
page 52

1 step = 31,000,000 miles
1 step = 31,000,000 miles
1 step = 31,000,000 miles
1½ step = 46,500,000 miles
10 ½ steps = 325,500,000 miles
12 ½ steps = 387,500,000 miles
28 steps = 868,000,000 miles
31 ½ steps = 976,500,000 miles
27 ½ steps = 852,500,000 miles

Ozone Destroyers
page 54

1. 73.1% increase
2. 27.1% decrease
3. 178.7% increase
4. 45.6% decrease
5. 94.7% decrease
6. 28.2% increase
7. 63.4% decrease

Piano Task Card
page 56

1. $F = 220\ (2^{12/12}) = 220\ (2) = 440$
2. $F = 220\ 2^{3/12}) = 220\ (2^{1/4}) = 261.6$
3. $F = 220\ (2^{15/12}) = 220\ (2^{5/4}) = 523.2$
4. The higher note has twice the frequency.

Hot Potato **page 58**

3. $T = \dfrac{40\,(237 - (212 - 1.85x))}{(212 - 1.85x) - 162}$

4. $T = \dfrac{40\,(25 + 1.85x)}{(50 - 1.85x)}$

5. 4.945 ft.	33.4 min.
6. 0.021 thousand ft.	20 min.
7. 6.9 thousand ft.	40.6 min.
8. 4.155 thousand ft.	30.9 min.
9. 2.03 thousand ft.	24.9 min.
10. 1.04 thousand ft.	22.4 min.
11. Las Vegas and Omaha	2.5 min.

Mod-Math Quilt Task Card **page 60**

1.
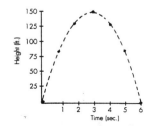

2. Answers will vary.
3. Answers will vary.

Tessellating Quilt Task Card **page 62**

Designs will vary.

Up and Down the Dial **page 64**

1. a. 0.06
   b. 5,000,000,000
   c. 5 GHz
2. a. w = 300,000,000/f
   b. 7,250,000
   c. 41.38 m
3. a. 300 MHz to 30 MHz
   b. 1m to 0.1m

Sound Off **page 66**

1. 0.02 sec.
2. 64 Hz
3. Decreases
4. 1
5. 1,080 ft./sec.
6. 2.45 ft.
7. 4,400 Hz
8. About 0.01 sec.

Sun and Moon **page 68**

1. 880,000 mi., 440,000 mi.
2. 2,160 mi., 1,080 mi.
3. Proportions may vary.
   Equivalent to MX/EM = SY/ES
4. 228,000 mi.
5. 201,000,000 sq mi.
   134,000,000 sq mi.

What Goes Up Must Come Down **Page 70**

1. a. 144
   b. Possible: It starts to fall.
   c. no
   d. Possible: same height going up and coming down

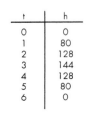

t	h
0	0
1	80
2	128
3	144
4	128
5	80
6	0

2. a. 2,304
   b. 2,288
   c. 2,288
   d. Possible: same height going up and coming down
3. a. 1,008; 1,024; 1,008
   b. highest point
   c. 0
   d. The rock hit the ground.

Uranium Decay **page 72**

1.
Half-life Unit	Unstable Atoms
0	100
1	50
2	25
3	12.5
4	6.25
5	3.125

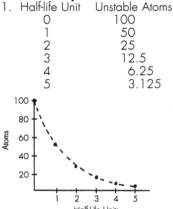

2. 1,408 million years or 1.4 billion years (2 half-lives results in ¼ of the radioactive elements)

3.
Days	Bacteria Cells
1	100
2	200
3	400
4	800
5	1,600
6	3,200

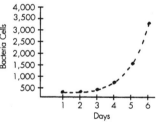

4. Yesterday 5,000   Tomorrow 20,000

Virtuoso Violins **page 74**

1. [graph: fingerboard (in.) vs. neck (in.)]
2. [graph: total length (in.) vs. body (in.)]

3. approx. 1.4
4. approx. 1.6
5. Violin 1.32
   Viola 1.32
   Cello 1.43
   Bass 1.4
6. Violin 1.65
   Viola 1.65
   Cello 1.63
   Bass 1.64

Ride the Wave   page 76

Across

1. 35
2. 1,140
4. 0.58
6. 49.4
8. 2.5

Down

1. 30
2. 104
3. 42.5
5. 80.4
7. 9.75

Blowing in the Wind   page 78

Scale Number	Wind (mph)
0	0
1	3
2	7
3	11
4	16
5	21
6	27
7	33
8	40
9	47
10	55
11	63
12	72

FRANCIS BEAUFORT

Copycat   page 80

1. E
2. O
3. R
4. P
5. Y
6. H
7. A
8. G
9. R
10. X

XEROGRAPHY

X-Ray Vision   page 82

1. 1,000 XU (L)
2. 20,000 (ST)
3. 10^4 (LI)
4. 1895 (TH)
5. 10^{-4} (A)
6. 10^{-8} (ER)
7. 10^{-11} (RO)
8. 20 (S)

WILHELM ROENTGEN

Days: Past, Present, and Future   page 84

1. GO—Tues.
2. TH—Tues.
3. AL—Sat.
4. M—Wed.
5. RI—Wed.

Bonus: Answers will vary.

ALGORITHM

Absolutely Zero Task Card   page 86

1-3.

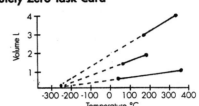

4. Between -250°C and -300°C, about -275°C
5. Student choice
6. -273°C

It's a Zoo Out There   page 88

1. 240
2. 24
3. 15
4. 54
5. 88
6. 1,760
7. 300
8. 4
9. 60

"IT DOES NOT DRINK"